SOI CIRCUIT DESIGN CONCEPTS

SOI CIRCUIT DESIGN CONCEPTS

Kerry Bernstein
IBM Microelectronics

and

Norman J. Rohrer
IBM Microelectronics

KLUWER ACADEMIC PUBLISHERS
Boston / Dordrecht / London

Distributors for North, Central and South America:
Kluwer Academic Publishers
101 Philip Drive, Assinippi Park
Norwell, Massachusetts 02061 USA
Telephone (781) 871-6600
Fax (781) 871-6528
E-Mail <kluwer@wkap.com>

Distributors for all other countries:
Kluwer Academic Publishers Group
Distribution Centre
Post Office Box 322
3300 AH Dordrecht, THE NETHERLANDS
Telephone 31 78 6392 392
Fax 31 78 6546 474
E-Mail <orderdept@wkap.nl>

Electronic Services <http://www.wkap.nl>

Library of Congress Cataloging-in-Publication Data
Bernstein, Kerry, 1956-
  SOI circuit design concepts / Kerry Bernstein and Norman J. Rohrer.
    p. cm.
  Includes bibliographical references and index.
  ISBN: 0-7923-7762-1 (alk. paper)
  1. Metal oxide semiconductors, Complementary--Design and construction. 2.
Silicon-on-insulator technology. I. Rohrer, Norman J. II. Title

  TK7871.99.M44B48 2000
  621.39'732--dc21   99-088175

Copyright © 2000 by Kluwer Academic Publishers, Second Printing 2000.

Printed on acid-free paper.

Printed in the United States of America

For Denise: As time goes by, I see how lucky I am.

- K. B.

To: Nathan and Luke,

May your curiosity lead you down a path full of wondrous endeavors and may your search for knowledge be meaningful and rewarding.

- N. J. R.

TABLE OF CONTENTS

Preface

In Igor Stravinsky's ballet "The Firebird", Prince Ivan captures the exotic Firebird in the magic garden of the sorcerer Kashchei. The prince releases Firebird when she gives him a feather which can be used to summon her in times of danger. The Prince falls in love with a princess held by the evil Kashchei, and eventually defeats the sorcerer and wins the princess' freedom with the Firebird's help.

It is fitting that much of this text was written to the music of this ballet. At once elegant, graceful, coordinated; and disjointed, unexpected, and misunderstood, "The Firebird" all too closely resembles the nuances of Silicon-on-Insulator technology.

As we complete this textbook, it is finally becoming clear that our industry no longer regards silicon-on-insulator technology as being in the same category as gravity-powered shoe air-conditioners and jet-powered surfboards[1]. SOI has been incubating for perhaps 20 years. Although the pressures of scaling have increased the urgency of its development, there are those who argue that it's behavior remains too unpredictable to be accurately modeled. Suffice it to say that SOI's potential performance rewards for enduring this complexity are substantial. There is no question that SOI is not an easy technology to learn and design with; using the technology requires a thorough, fundamental comprehension, and demands considerably more designer mental energy than its bulk CMOS predecessor did. The reader must remember, however, that

---

1. See "The Gallery of Obscure Patents" at *http://www.patents.ibm.com/gallery* on the World Wide Web.

although SOI might be difficult, conventional CMOS was also originally referred to as the S.O.B. (Silicon-On-Bulk, of course!) technology when first introduced.

Chapter 1 provides the reader with a brief review of how CMOS technology has been scaled through multiple lithography generations, and describes the device physics which threaten conventional CMOS' future. The case for SOI is built as a means to continue improvement in density, performance, power and integration.

Chapter 2 introduces the basic SOI physical device structure, and how it differs from its bulk CMOS predecessor. Manufacturing considerations and materials properties are highlighted in this section.

Chapter 3 addresses the fundamental electrical properties associated with the SOI device, and how the structures described in the previous chapter give rise to these properties. If the reader wishes to "cut to the chase," this chapter and the following two provide the most essential of SOI core concepts.

Chapter 4 explores the response of static circuits to SOI technology and offers guidance on static design practice.

Similarly, Chapter 5 provides an introduction to the issues associated with dynamic circuits in SOI, and includes design practices disclosed in the recent literature which mitigate many of the undesirable dynamic problems.

SRAM and cache arrays are probably the hardest designs to move into SOI. Chapter 6 is entirely devoted to addressing the issues surrounding SOI SRAMs. A few suggestions are offered on how to get around some of these problems.

In addition to the SRAM, selected other special function circuits require some finesse when migrated into SOI, and are described in Chapter 7. Off-chip-drivers, phase-locked-loops, and Electro-static Discharge devices are among these unique structures.

Chapter 8 addresses global design considerations which are unique to SOI. Power distribution, clock branching, and noise is discussed.

Finally, Chapter 9 offers a glimpse into some of the future leverage SOI may offer. Rather than designing to avoid SOI's idiosyncrasies, the true benefit of the technology may be realized when these features are exploited.

Acknowledgments

This text draws heavily from the knowledge base in the literature, as well as from the author's direct experience in PD-SOI at IBM Microelectronics. SOI's emergence as a mainstream technology within IBM is clearly due to the persistence of Ghavam Sha-hidi and Fari Assaderaghi, and their unwavering belief in this technology's merits and timeliness. Ghavam and Fari were the first sources of SOI device understanding within the company. They initiated the rest of IBM's device and circuit design com-munities on the learning curve for SOI behavior. The authors are grateful for their guidance and foresight.

Fulfilling the commitment of creating a textbook on a new technology is also impos-sible without the implicit approval and support of one's family and supervisors; Kerry wishes to thank Hallie, Risa, and Mikey Bernstein for letting their dad get this book out of his system. You guys are the air that I breathe - I love you, goobers. Norman wishes to thank his wife, Melinda, for the support to work on this book many eve-nings. Her support extends well beyond the arena of another book. The authors also recognize their management team for their patience; Chekib Akrout in Austin, TX, Henry Levine, Sol Lewin and Scottie Ginn in Essex Junction, VT, and Bijan Davari in East Fishkill, NY.

The authors are quite indebted to the following individuals for valuable concepts, inputs and conversations: Dean Adams, David Allen, Ingo Aller, Tony Aippersbach, Ron Bolam, Andy Bryant, Miles Canada, Kent Chuang, Mike Ciraula, Dennis Cox, Emmanuel Crabbe, Andy Davies, Dan Dreps, Patrick Hansen, Bill Klaasen, Karl Kroell, Ed Nowak, Nghia Phan, Don Plass, Devandra Sadana, John Sheets, Melanie Sherony, Jeff Sleight, Sam Storino, Yuan Taur, Steve Voldman, Larry Wagner, Lynn Warriner, and Isabel Yang. Original excerpts from many of their seminal works appear in this text. Finally, Howard Gieselman, Ione Minot, and Keith Williams are recognized for their timely support.

Kerry Bernstein
Norman J. Rohrer

Underhill, Vermont
November, 1999

CHAPTER 1

The Time for SOI

1.1 Technology Scaling in VLSI

Very Large Scale Integration (VLSI) employing bulk-substrate CMOS technology is truly the industrial success story of the twentieth century. In a relatively brief period of time, CMOS has not only replaced bipolar transistors in most commercial applications, it has defined whole new markets and applications. The engine behind this phenomena, *Technology Scaling*, has fueled relentless growth in semiconductor performance and density.

Scaling refers to an observation made by Bob Dennard, et al, in 1974 that the features of an idealized transistor and interconnect, when migrated from one technology generation to the next, may all be translated by a common scaling factor [1.1]. Observing this scale factor then maintains established optimum electrical properties.

> **Rule of thumb:** Historically, scaling factors of 1.25 to 1.33 have been realized in moving to next-generation technologies.

Figure 1.1 depicts a number of these physical device parameters in a common N-type MOSFET in a "common substrate"[1], and their relationship to scaling factor a. In an idealized setting, scaling to the next generation technology has enabled density, performance, and power to improve by a factor of a^2, a, and $(1/a^2)$, respectively.

---

1. Or *bulk substrate*, as referred to in this text.

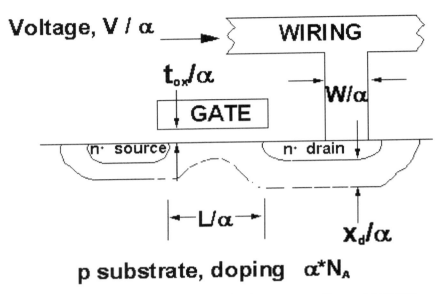

FIGURE 1.1 Ideal scaling by factor a practiced on common N-type MOSFET

As a result, CMOS processor performance has been increasing at a rate predicted by Moore's Law, 2X every 18 months. Figure 1.2 graphically displays this story over the past 20 years. Product designers leveraged this improved performance and density by integrating function previously done in separate components on board the processor. Because of this, die size and chip power have not been decreasing rapidly.

1.2 The End of Moore's Law?

Time has taken its toll on Moore's law. Figure 1.3 shows the original fit to design points from the seventies [1.2] and a return by its progenitor to the observation in 1995 [1.3]. The reduced gain per generation is attributed to a number of factors associated with increased complexity. The departure from this idealized fit is observed to be getting worse. A more profound limitation, capable of dramatically impacting Moore's Law and scaling, has emerged more recently, however, and is the motivation for the development of SOI technology. So please read on!

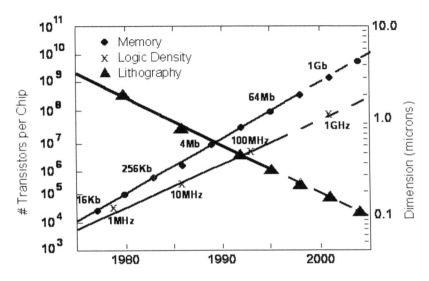

FIGURE 1.2 Memory and logic leveraging of device count and lithography improvements in VLSI over the past 20 years

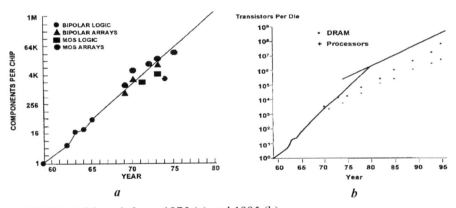

FIGURE 1.3 Moore's Law: 1975 (a) and 1995 (b).

The predominant boundary to scaling was determined by limitations found in the fabrication tools (i.e. minimum feature size, photolithography) and materials of the current technology (i.e gate dielectrics). It is now becoming apparent that MOSFET device threshold voltages may no longer be reduced much more, due to

(a) *excessive subthreshold leakage impacting circuit function,*

(b) *precarious noise immunity,*

(c) *non-conducting device wearout limitations, and*

(d) *chip standby power.*

To counter the resulting net loss of overdrive by not reducing threshold voltage, V_{DD} in turn has not been reduced by the full scaling factor. As physical dimensions of the device have continued to scale, however, the growing lateral electric field in the MOSFET has degraded device mobilities and hence the performance gain per generation, as shown in Figure 1.4 [1.4].

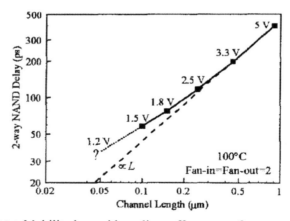

FIGURE 1.4 Mobility loss with scaling; effect on performance [1.4].

Given this limitation to threshold voltage reduction, it is obvious that in the absence of the introduction of new technologies, CMOS performance will continue to flatten or may actually degrade. Figure 1.5 shows actual inverter delays reported in the literature plotted with a prediction of future bulk CMOS performance [1.5]. Various predictions have anticipated the useful lifetime of the present CMOS device technology paradigm. Credible arguments anticipate performance improvements are severely curtailed by the 100 nm lithography generation [1.5]. The resulting erosion of performance in the microprocessor would soon become quite pronounced, in lieu of a device alteration. Figure 1.6 shows expected trends in microprocessor cycle time in the not-too-distant future.[1.6]

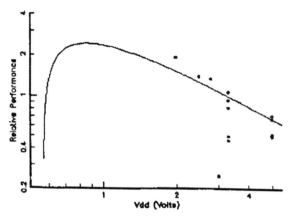

FIGURE 1.5 Bulk CMOS vs. inverter performance with scaling. [1.5]

Evolution of Processor Clock Frequency

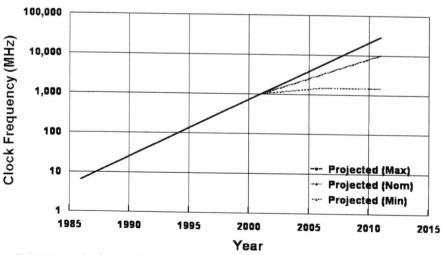

FIGURE 1.6 Anticipated processor response to future CMOS device design points.

Students of VLSI may argue that soothsayers have been predicting the end of Moore's Law every year since its inception; while somehow the industry has continued to find solutions, the coming generation's challenges are more ominous.

Works in the literature have proposed technologies deliberately accepting high leakage. It has been shown that CMOS may be operated very efficiently when a transistor's I_{ON} is equal to I_{OFF}, occurring at approximately 400 mV [1.7]. This novel approach to logic computation requires an infrastructure and a paradigm change which industry has not yet displayed a willingness to accept.

With the chip designer being forced to accept painful characteristics to continue improving performance conventionally, it was clear that new transistor technologies would be needed, as V_{DD} approaches 1.0 V, in order to continue. Unlike the transition from bipolar to CMOS transistors, no fundamentally different semiconductor device is waiting in the wings to replace CMOS. It follows, then, that alterations which enhance the field-effect transistor's transconductance while leveraging existing design tools and methodologies would be of great interest. SOI technology matches this description quite nicely.

1.3 The Case for PD-SOI

1.3.1 Why Partially Depleted SOI?

The motivation for moving to SOI clearly is to extend scalability of CMOS, to continue enjoying performance as well as density benefits of migrating designs and architectures to next generation lithographies. It makes sense then, that the following would be first order considerations in selecting a transistor device technology:

1. Process Commonality
   PD-SOI may be thought of as a *evolutionary* rather than *revolutionary* device structure, in that with the exception of the wafer SIMOX formation, the CMOS wafer fabrication is performed on the same bulk CMOS tool set, targeting parameters which are either similar or identical to bulk CMOS.

2. Scalability
   A device definition which, when scaled through successive lithographies, continues to produce a viable product is very important. Because PD-SOI does not require the ultra-thin active silicon thicknesses typically used in fully-depleted (FD) SOI, investments in its developed can be returned over more generations

3. Tolerance

Using identical materials, tools, processes, and parameter targets as in its bulk predecessor, the spatial variability in SOI electrical parameters is well established and accommodated in the software tools commonly used. That leaves the use-dependent variation, which can be shown to be well modeled.

With FD-SOI, the entire body of the transistor is inverted. The threshold voltage, therefore, is a function of the charge contained in the body, and can vary substantially across chip. As active silicon thickness is driven thinner, this becomes a boundary condition [1.8]. The reader is referred to the literature for more details on FD-SOI [1.9].

4. Few emerging alternatives exist

   CMOS LSI was developed as bipolar technologies were being used in most high speed applications. When CMOS matured enough to be viable based on performance, power and cost, the transition was made. Unlike those times, there is no new paradigm in transistor electronics waiting "in the wings." Until an entirely new successor is established, our industry will continue to introduce innovative CMOS variations such as SOI into volume production.

1.3.2 Why SOI *Now*?

Oddly enough, there is nothing new about Silicon-on-Insulator (SOI) technology. It's been around for years! Figure 1.7 summarizes traditional SOI products in the electronics marketplace. Active silicon thickness, a parameter described in the next chap-

Market	Si Thickness	Applications
Space	$\sim 10\,\mu$	MEMS
		Guidance/Gyros
Power		
	$\sim 2\,\mu$	IGFETs
		Accelerometers
Analog	$\sim 2\,\mu$	Auto Electronics
		Cross-Bar Switches
CMOS Logic	~ 200 nm ($0.2\,\mu$ and below)	Microprocessors
		Microcontrollers

FIGURE 1.7 SOI uses in the existing marketplace.

ter, is characteristic of the application, as shown in Figure 1.7. It becomes clear that SOI applications exploit specific attributes of the device; these specific applications employ distinctly different active silicon thicknesses. SOI use in high speed microprocessor applications has been unattractive up to now, however, due to discouraging SOI device behavior at active silicon thicknesses necessary for high performance:

- The means to manufacture SOI wafers inexpensively and relatively defect-free was elusive.
- The Floating Body Effect of Partially Depleted SOI MOSFETs scared away a design community which was used to dealing with predominantly spatial rather than temporal device variation. (Don't worry; we'll cover this notion in much greater detail in the following pages.)

So what has changed, given that the VLSI industry is showing re-awakened interest in SOI technology?

- First, the means of fabricating SOI wafers has matured. Defect density in the silicon caused during formation of the prerequisite buried oxide is accommodated in the device's operation as we will see in Chapter 2.
- Second, circuit design using the SOI MOSFET has climbed the learning curve. The History Effect of the SOI device was an unattractive characteristic which scared the design community away from its use. It has been demonstrated that, in fact, SOI's history-dependent device behavior in most cases can be harnessed, and that impressive designs may be realized [1.10] [1.11]. Given the substantial performance improvement of the SOI technology, modifying the necessary design conventions is a good deal.
- Power dissipation density is becoming a first order design consideration. Traditionally a concern for portable electronic applications, increased integration in desktop, server, and mainframe engines has driven die power to the limits of our ability to cool it. With its decreased junction capacitance and improved performance at a given voltage, SOI is often selected on its power merits alone.
- Scaling has drastically reduced both good and bad on-chip capacitances, leaving even the logic on recent generation hardware vulnerable to radiation-induced soft errors. SOI helps to address this concern, which otherwise discourages the application of VLSI CMOS in satellite and aerospace platforms.
- Finally, a trend towards building more analog. mixed signal, DSP, and RF function on chip has been noted. Substrate-coupled noise has been a persistent bulk-substrate problem, however, which has limited the integration of these CMOS extensions. SOI's intrinsic body isolation is well suited to providing the low noise foundation which accommodates both high-noise-generating static combinatorial CMOS and low-noise-tolerant analog signal processing. SOI is one venue which will encourage continued merging of these functions on a single die.

As we will repeat many times in this text, SOI achieves its performance advantage over conventional bulk CMOS from 3 intrinsic features:

1. Less Junction Capacitance

    Rule of thumb: Junction capacitance reduction is responsible for approximately half of Partially-Depleted SOI's advantage

2. Lower average device threshold voltages

    Rule of thumb: SOI's lower threshold voltage provides the other half of SOI's performance advantage.

3. Less body effect and source follower effect

    Rule of thumb: Lower source-body bias produces limited performance advantages in SOI, in typical 2-high bulk configurations. Higher device stack performance will have substantial improvement due to body effect.

If the boundary conditions mentioned previously in fact eventually assert the lower limit for future scaled technologies, then the advantage provided by #1 (above) will continue to boost performance, even if the other two do not due to minimum threshold voltage. In Chapter 9, we will explore the dynamic threshold feature, another future opportunity which SOI enables. Dynamic threshold variation can free the designer from living with the I_{ON} and I_{OFF} characteristics of a MOSFET with a single threshold voltage.

Dynamic threshold variation is but one example of how the technology's idiosyncrasies might be exploited rather than accommodated in the future. Alas, much of the experience shared in this text is associated with ways to perpetuate existing bulk circuit structures and design practices in the SOI technology.[2] Following these first tentative steps, the next phase that PD-SOI CMOS will assume will be to harness SOI history effects in order to continue finding innovative paths to microprocessor performance improvement. The authors look forward to writing that book when the industry learns how to take this second step.

1.4 Summary

It is becoming evident that continued CMOS scaling into the deep submicron region is causing the limits of bulk substrate MOSFETs to emerge. With diminishing perfor-

---

2. The first generation of products built in SOI have assumed this conservative approach. Minimizing risk means limiting exposure to multiple new technologies.

mance and higher leakage currents in store for the coming lithography generations, Partially Depleted SOI, which had been used in other markets and applications, is now becoming a more attractive high speed technology alternative. Advances in processing and fabrication of SIMOX wafers, as well as a better understanding of SOI circuit behavior make this technology a candidate for volume production in the present generation.

SOI also has known design liabilities which, because the more simple bulk CMOS option provided acceptable power and delay, discouraged its past use. With the substantial advantages SOI is now noted for, it has become worthwhile to examine more closely its electrical behavior, so that robust, complex SOI product designs may be initiated. Continued learning of preferred circuit design practice from recent experience has developed a body of SOI circuit design knowledge, which is captured for the reader in the balance of this text.

REFERENCES

[1.1]   R. H. Dennard, et al., "Design of ion implanted MOSFETs with very small physical dimensions," *IEEE Journal of Solid State Circuits,* Vol. SC-9, October 1974, pp 256-268.

[1.2]   G. E. Moore, "Progress in Digital Integrated Electronics," *Proceedings, IEEE International Electron Devices Meeting,* pp. 11-13, 1975.

[1.3]   G.E. Moore, "Lithography and the FUture of Moore's Law," *Proceedings of The Society of Photo-Optical Instrumentation Engineers (SPIE) Conference,* 20-21 February 1995, Santa Clara, CA, Vol. 2437, pp. 2-17.

[1.4]   Y. Taur, et al., "CMOS Devices below 0.1 μm: How High Will Performance Go?" *Proceedings of IEEE International Electron Devices Meeting,* 1997, pp. 215-218.

[1.5]   E. J. Nowak, "Ultimate CMOS ULSI Performance," *Proceedings of IEEE Electron Devices Meeting,* 1993, pp. 115-118.

[1.6]   R. D. Isaac, "Beyond Gigahertz and Gigabit Chips," *IBM Corporation, Internal Presentation.*

[1.7]   A. J. Bhavnagarwala, et al., "Projections for High Performance, Minimum Power CMOS ASIC Technologies, 1998-2010," *Proceedings of 10th Annual IEEE International ASIC Conference and Exhibit,* pp. 185-188, 1997.

[1.8]   N. Kistler, et al., "Scaling Behavior of Sub-Micron MOSFETs on Fully Depleted SOI," *Solid-State Electronics,* Vol 39, No. 4, pp. 445-454, 1996.

[1.9]   J. A. Burns, et al., "Design Criteria for a Fully-Depleted 0.1 μm SOI Technology," *Proceedings of 1997 IEEE International SOI Conference,* pp.78-79, October, 1997.

[1.10] M. G. Canada, et al., "A 580 MHz 32b PowerPC Microprocessor," *Proceedings of IEEE International Solid State Circuits Conference,* pp. 430-431, February, 1999.

[1.11] D. H. Allen, et al., "A 0.2μm 1.8V SOI 550MHz 64b PowerPC Microprocessor with Copper Interconnects," *Proceedings of 1999 IEEE International Solid State Circuits Conference*, February, 1999, pp. 438-439.

CHAPTER 2   *SOI Device Structures*

2.1 Introduction

Silicon On Insulator (SOI) structures do not vary much from normal bulk CMOS. The major difference is the insertion of the insulation layer beneath the devices. Once this is accomplished, one could continue to use the identical bulk CMOS process and fabricate the devices. No changes to the process would be required, however, to take full advantage of SOI; small changes to the process are required. This chapter will discuss the physical structures and fabrication techniques of the wafer, FETs, diodes, resistors and thin oxide capacitors. Intertwined with the physical descriptions will be the mention of process changes to enhance the SOI device's usability and performance.

Another process for isolating FETs from each other is silicon on sapphire (SOS). This technique used a sapphire substrate and etched small mesas of silicon on its surface. The FETs on SOS can be made similar to the FETs on SOI. This technique was expensive to use since it required sapphire substrates and was not very manufacturable. The methods described in this chapter will use techniques that are compatible with today's manufacturing process for bulk CMOS wafers.

The subject of this chapter is the basic structure of devices found in SOI CMOS technologies. This includes the wafer structure, FET cross-sections, diode profiles, decoupling capacitor structures and resistor topologies.

2.2 Wafer Fabrication

<u>So how do I get started?</u>

The objective is to create a insulating layer beneath the devices such that the body of the FETs are not connected when the shallow trench isolation (STI) is used to pattern the active silicon areas. This insulating region is typically made of silicon dioxide and is formed one of three techniques: SIMOX, Bonded SOI or Smart Cut. This insulating region will be called the buried oxide or BOX.

2.2.1 SIMOX

SIMOX stands for the Separation by the IMplantation of OXygen. In this technique, an oxygen implant is performed on an epitaxial wafer before it has started any other CMOS processing.

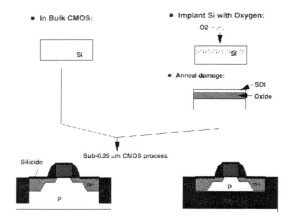

FIGURE 2.1 SIMOX process steps

The high energy implant forces a large dose of oxygen deep beneath the surface. Once the implant is complete, a high temperature anneal is done to form the silicon dioxide insulating region. The anneal must also help to recrystallize the silicon in the epitaxial layer which is at the surface of the wafer. Due to the high energy and high dosage of the implant, the surface of the wafer has been heavily damaged and silicon atoms have been pushed deeper into the wafer causing a surplus of Si atoms beneath the surface. The post-implant anneal must recrystallize the surface to produce high quality silicon for the subsequent CMOS processing. During the recrystallization anneal, the volume expansion of the oxygen converting to SiO_2 results in additional mechanical stress to accumulate at the Si/SiO_2 interface. This may result in forming silicon dislocations or pipes that are 0.2 to 1.0μm long, see Figure 2.2. These dislocations are long

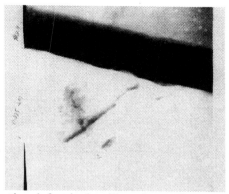

FIGURE 2.2 Dislocation defect in the silicon layer after recrystallization anneal.

enough to create a leakage path that may from source to the drain of an FET if one lines up with the length of the FET.

The dose of the implant required is dependent upon the device design point that is required for the given product. Too large of a dose is slow and expensive and will cause more defects in the top layer of silicon. Too small of a dose will not reduce the junction capacitance enough and will not electrically remove the silicon beneath the buried oxide from the active device. A typical dose for SIMOX is 10^{16} to $5e10^{17}$ cm^{-3}[2.7]. Therein, the buried oxide may be considered another FET oxide with the backside silicon as the gate. This will be discussed more in Chapter 8.

The buried oxide varies in thickness from 5nm [2.1] to 400nm [2.5] with varying doping. The energy of the implant will determine the silicon thickness above the buried oxide. Thicknesses of 50 nm [2.2] to 180 nm [2.6] are used for typical microprocessor designs. Other applications, such as space applications or power devices, will use much thicker active silicon regions. The thickness of the silicon layer is more related to the design of the FET. If one chooses a partially depleted FET instead of a fully depleted FET, the silicon layer is going to be thicker.

As with any added process step, the creation of the BOX is not for free. It requires a high energy implanter than can supply a very large dose of oxygen. Currently, about 20-40 wafers per day are created with this technique on a single implanter. As lower doses are applied, this number will increase, but this step is still a time consuming process. Another artifact of the implantation is that not all of the silicon is converted to SiO_2. When this occurs, small locations of silicon reside in the BOX. Due to the high dose, the occurrence of silicon residuals is most likely to occur at the bottom of the BOX and not near the device. These residuals are called inclusions.

2.2.2 Bonded SOI Wafers

Bonded SOI wafers create the buried oxide without ion implantation. Figure 2.3

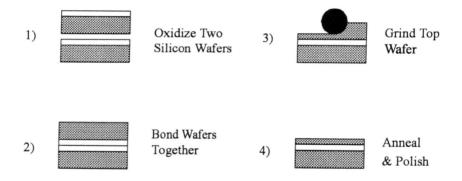

FIGURE 2.3 Process steps for SOI wafers created using the bonded wafer technique.

shows the processing steps for bonded wafers. First, two silicon wafers are heated to form a silicon dioxide layer on the surface. The two oxide surfaces are then bonded together to form the buried oxide. The backside of one of the silicon wafers is ground down to the desired thickness. Finally the SOI wafer is annealed and polished to leave a thin layer of silicon above the buried oxide that is electronic grade quality[2.7]. This wafer is then ready for typical CMOS processing. One problem with bonded wafers is that it required the use of two silicon wafers to provide one SOI wafer.

2.2.3 Smart Cut

The final technique discussed here is called Smart Cut. Figure 2.4 shows the steps in

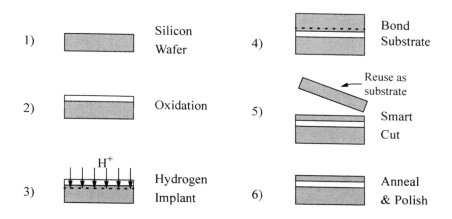

FIGURE 2.4 Process steps for Smart Cut.

the smart cut process. This technique starts with the oxidation of a silicon wafer. The wafer is implanted with protons (hydrogen nucleus) through the oxide layer. The surface of the oxide is clean. A second silicon wafer is bonded to the oxide layer (step 4). This wafer is now the substrate for the SOI wafer. A thermal anneal creates a stress fracture along the plane of the hydrogen implant. This is the Smart Cut. The original silicon wafer can be removed from the trilayer stack leaving behind a thin layer of silicon on the top of the buried oxide. The removed portion of the silicon wafer will become the substrate for another smart cut wafer. In this manner, no silicon is wasted. To finish process, the SOI wafer is annealed and polished to prepare the surface for traditional CMOS processing steps.

2.3 Patterning SOI regions

What if I only want some SOI circuits?

It is possible to pattern the region that is dedicated to bulk CMOS separate from the region that will be SOI. The patterning techniques are most easily applied to the

SIMOX process since one would be patterning the implantation region. Both bonded SOI and smart cut would require an alignment of the patterned areas to each other. The transition region between native bulk silicon and SOI regions is dependent upon the thickness of the buried oxide. This transition region is not usable for device processing due to the stress that exists in this region. Figure 2.5 shows an example of a transition region. The addition of the oxygen from the implant and the subsequent anneal creates a small thickness difference between the bulk and SOI region. This thickness difference is highly stressed and can result in cracking of the silicon at the surface (See Figure 2.5). Transition regions are in the tens of micrometers range for a

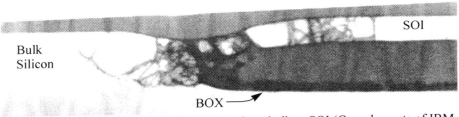

FIGURE 2.5 TEM of the transition region from bulk to SOI (Complements of IBM Research).

2000 μm thick buried oxides. As the buried oxide thickness become smaller, the transition region can become less than 5 μm. Therefore, if one desires to have bulk circuits on one part of the chip die and SOI circuits on another part of the die, it is feasible, though not regularly practiced.

One should not take too lightly the small step that occurs in the transition region. This height difference can be a real challenge for photolithography, trying to pattern the next several levels. The focus depth of today's lenses are very shallow and cannot focus at two different depths for the bulk and SOI portions of the wafer.

2.4 Transistor Structures

<u>I have a buried oxide, now what?</u>

The MOSFET is the structure under scrutiny and research in both bulk CMOS and SOI CMOS. In bulk, there is only one main device structure and the challenge presented to device engineers is how to make the fastest, most reliable device. The choice of thresholds voltages, junction depths and background doping levels all have driven scaling of the FET to the to ever-finer device dimensions.

With SOI, a device engineer needs to make one more basic decision before selecting the rest of the device profiles and parameter. The question to be answered is: Are you going to work with a partially depleted device or a fully depleted device. As with most engineering questions, this one dictates or is dictated by other manufacturability questions such as control of the threshold voltage, the silicon thickness, etc. Let's look at the device structures.

2.4.1 Partially Depleted FET Structure

Keep your body in control!

Relying upon one's understanding of a bulk CMOS FET, the SOI FET has many features in common. Figure 2.6 shows the cross-section of the partially depleted SOI

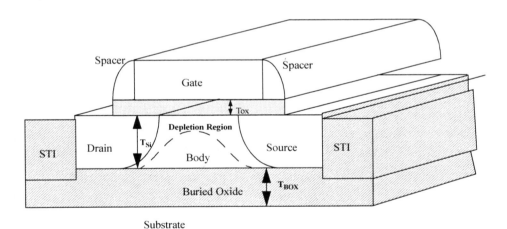

FIGURE 2.6 Cross-section of a partially depleted SOI FET.

device. This figure is not drawn to scale. The silicon is insulated with silicon dioxide on all sides. Beneath the silicon is the buried oxide that was created before any device processing was started. To the left and the right of the active silicon, the shallow trench isolation (STI) is the same STI isolation technique that is used in bulk CMOS technologies. The shallow trench isolation is greater than the thickness of the silicon, T_{Si}, resulting in a continuous silicon dioxide region from the wafer surface through the buried oxide beside the FET. The gate oxide thickness, T_{ox}, is identical to a bulk technology as is the gate stack thickness and the spacer dimensions.[1]

The depletion region extends into the body of the FET under the gate at the source-body and drain-body junctions. It does not deplete all of the charge in the body, resulting in the name *Partially Depleted* SOI.

One cannot control the voltage of the body with out having some region of the body containing charge. With the partially depleted body, there is a high resistance across the body, but the charge is mobile. The series resistance down the length of the body is large and variable, as the depletion width is voltage dependent. As the depletion width from the source or drain into the body increases with voltage bias, the cross sectional area containing charge becomes smaller and the resistance increases.

This mobile charge results in unique SOI-only device and circuit characteristics that will be described at length in several subsequent chapters.

This structure has a parasitic bipolar device in parallel with the FET. For an NFET, the parasitic bipolar device is an NPN structure. Without a contact to the body, the body is floating and under DC conditions is driven to a potential determined by the leakage currents from the source and drain to the body across the diodes that form the junctions. Under AC conditions, the diodes between the body and the source or drain, the impact ionization currents from the FET and the capacitively coupled currents from the other three terminals determining the potential on the body. When the body is floating for the FET this also means that the base of the parasitic bipolar device is floating. One wants to minimize the potential in the base region to prevent the bipolar device from turning on. In a similar structure, there is a parasitic PNP bipolar transistor that is in parallel the a PFET. The bipolar devices do not have very high gain since the base of the NPN or the PNP is quite wide. When the body is floating, there is no direct connection to the base of the bipolar transistor to sustain the potential.

---

1. A fully depleted FET structure is very similar to the partially depleted structure. It has the same STI isolation and it has buried oxide beneath the silicon. The one big difference is that the silicon thickness of a fully depleted device is thinner than the silicon thickness of the partially depleted device. The background doping of the NWELL is lower than in partially depleted devices. This allows the depletion regions from the source to body and from the drain to body regions to fully deplete the body of mobile charge under all bias conditions. The silicon thickness of a fully depleted device is very critical. This thickness determines the threshold voltage of the device and is a manufacturability issue. This text deals with partially depleted devices. See [2.3] for more information on fully depleted devices.

To lower the potential on the body, the source and drain region implants are done with a higher dose or a lower depth to increase the leakage across the diodes. With a more abrupt junction, the leakage helps to control the potential on the body.

As one looks from drain to source of the NFET in Figure 2.6, the dopant type is n-p-n. This structure creates a parasitic bipolar transistor. The drain is the collector, the body is the base and the source is the emitter. The base of the bipolar device is not connected to any potential. It is floating just like the body is floating. The parasitic device must be modeled and accounted for in device design points.

Another parasitic device that is apparent in Figure 2.6 is another NFET with the substrate acting as the gate and the BOX acting as the insulating oxide for the MOSFET. This device will be called the backside device.

Figure 2.7 shows an SEM cross section of a SIMOX wafer with tungsten local inter-

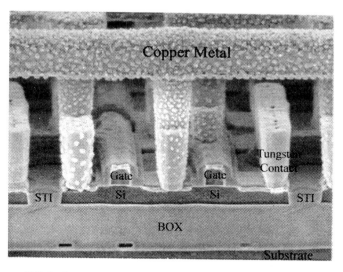

FIGURE 2.7 SEM of a cross section of two SOI FETs with tungsten contacts and copper wires.

connects, tungsten studs and one level of copper interconnects. The buried oxide (BOX) is much thicker than the SOI for this wafer. Here two FETs share a common silicon mesa and share the source connection between the two devices. The shallow trench isolation (STI) is present to the left and right of the devices and provides isolation from the other FETs nearby. The silicon above the buried oxide is surrounded by silicon dioxide on all sides. The only places where the silicon dioxide is not present is at the contacts to the source and drain of the device.

2.4.2 Body Contacts

How do I control the body?

In bulk CMOS, it is possible to have independent control of the body of each and every device. However, it requires the use of a "triple well" process and much more space than SOI. In a more traditional twin well CMOS process on an p-type substrate, the body's of the PFETs are controllable by controlling the potential of each respective NWELL. There is no control of the body of an individual NFET since they share the common substrate which is usually grounded.

In a partially depleted SOI CMOS FET structure, there is a conducting path beneath the FET that can be used to control the potential of the body. This requires a new structure in SOI since each silicon area is an isolated mesa from the rest of the substrate or NWELL. To provide the conducting path from the body itself to a contact to the body, one must create a p-type channel for an NFET and an n-type channel for the PFET.

In Figure 2.8 two types of body contact are shown for NFETs. In part (a) of the figure,

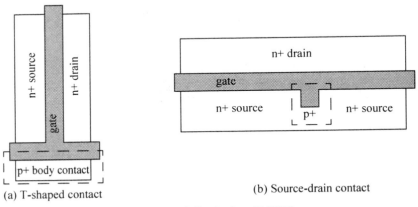

(a) T-shaped contact

(b) Source-drain contact

FIGURE 2.8 Body contacts for a partially depleted NFET.

the gate is a T-shape. A p+ region is created at the crossbar of the T-shaped gate. This p+ region is connected to the body of the NFET through the p-type region directly under the gate. Here, the body is a unique terminal and needs to be contacted with an interconnect and a via at the body contact to control the potential of the body. One consequence of the T-shaped connection is that the width of the NFET is now increased by a long channel length device that connects the source and drain under the crossbar portion of the gate. Also, the resistance down the body can be quite large, and requires that the device be contacted at both ends. In this case the polysilicon gate becomes an I-shape. One may also finger wide devices by forming a multiplicity of smaller parallel gates and using one body contact to control all of the fingers. Finally, two separate devices may share a common body by using two T-shaped gates and aligning the crossbars opposite the common body. In this case, the body contact can be connected to an external node, or the devices may share body potential, but the body is not driven to an external voltage. This is useful in a circuit that needs to have matched devices such as sense amps in SRAMs.

In part (b) of this figure, an alternate contact to the body is achieved. In this method, the p+ region is created in the same silicon as the n+ source, creating a butted junction. Once a salicide is created on the source, the salicide will short the p+ region to the n+ source. The p+ region will again connect to the body through the p-type region that is under the polysilicon gate. There is no need to connect this p+ region to an interconnect since it is already connected to the source of the NFET. One disadvantage of this contact is that one cannot choose another potential other than the source to control the body. In some applications, it is desirable to have independent control of the body of the FETs. In this case, the T-shaped body contact is the structure of choice.

Additional body contacts can be derived. For example, the T-shaped structure in Figure 2.8(a) can become a body-source short if the left side of the crossbar is removed. Then the T-shaped gate becomes an L-shaped gate and the source and body are shorted again by the salicide. This will alleviate some of the space requirements required with the T-shape structure. In any manner, body contacts are not widely used as they increase the amount of capacitance on the gate, increase the charge within the body and take up more silicon area. Only critical circuits will have body contacts on them.

PFETs have symmetrical structures for their body contacts, but the contact regions are now n+ and connect to the n-type silicon. This n-doped silicon region is all that remains of NWELL beneath the polysilicon gate.

2.5 Diodes

Oh where, Oh where did my other terminal go?

With transistors, the fourth terminal of the FET comes into play with SOI much more than bulk. However, with simple p/n diodes, the substrate and the NWELL have been cut off due to the buried oxide. The removal of the NWELL has lost the second terminal on a diode in a bulk diode as shown Figure 2.9(a). One may consider using a butted junction, but this would require an additional step to prevent the formation of a salicide over the butted junction. A different topology of the diode can be created with the same processing steps currently in use.

2.5.1 Poly Bounded or Gated Diodes

In SOI, the diode can be constructed again, a cross-section is shown in part (b) of

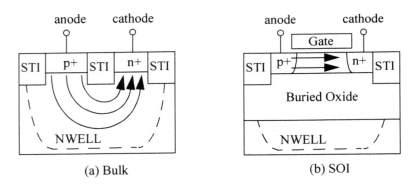

(a) Bulk   (b) SOI

FIGURE 2.9 Cross-section of a diode in bulk(a) and a poly bounded diode in SOI(b) with direction of current flow shown by the arrows.

Figure 2.9. The cross-section looks something like an FET; however the diffusions that would normally be the source and the drain of the FET are now the anode and the cathode of the diode. The anode is p+ and the cathode is n+. The body of this structure could either be an NWELL (as shown), or the p- epi-layer. In this cross-section, the p/n junction is at the anode/body junction. The diode is now a perimeter diode, where the bulk diode was an area diode, meaning that the current flow was though the center of the diode. In the poly bounded diode, the current is only flowing through the edge of the diode. To pass an identical current through a forward biased diode, the space required to layout a poly-bounded in SOI requires a larger space than a conventional diode in bulk CMOS.

The voltage on the gate determines the current carrying capability of the diode. One technique for using a gated diode is to force the gate to be connected to the cathode. This technique allows for the gate voltage to be maintained such that it does not effect the diodes current carrying capabilities. One should note that if the diode is used for an electro-static discharge protect device, (ESD Diode), on a multivoltage chip, the voltage of the output driver may cause the gate of the diode to exceed the reliability tolerance of the thin oxide. In this case, the gate should be driven to an intermediate voltage level that maintains the reliability of the oxide.

One complication of the poly bounded diode on SOI is that the series resistance is larger. The cross sectional area of the poly-bounded diode's NWELL is much thinner than the bulk diode's NWELL. This leads to series-limited current carrying capabilities of the diode. To correct for this, a much larger diode is used to reduce the total series resistance.

2.6 Resistors

Where is the path of least resistance?

The buried resistor and the polysilicon resistor are available in bulk as they are in SOI. The buried resistor is a depletion MOS device with its gate tied to the source. The resistance is determined by the channel region of the device, which remains unchanged. This results in the same resistance as well as the same tolerance in bulk and SOI technologies.

The polysilicon resistors are unchanged in SOI. As in bulk, they may be converted to silicide or left as polysilicon.

The NWELL resistor that is used in the bulk technologies is not operational in SOI since the buried oxide removes the silicon connection between two adjacent NWELL contacts. There is no conducting path through the NWELL in SOI. One of the other resistor structures must be used to replace NWELL resistors.

2.7 Decoupling Capacitors

I need charge.

The decoupling capacitor structure is nearly identical on SOI as it is for bulk CMOS. An accumulation capacitor, an NFET in an NWELL, is a simple structure for providing capacitance using the thin oxide of the FET. In this configuration, the source and drain are grounded, and hence the NWELL is grounded and the gate is tied to V_{DD}. As with the poly-bounded diode, the resistance of the body is large for the SOI decoupling capacitor and this results in the large RC component and reduces the capacitors frequency response. To alleviate this series resistance, the distance between the source and drain of the NFET should be kept short. This is a less-efficient use of space since the "length" of the FET is shorter than the bulk counterpart. Figure 2.10 shows the bulk and SOI decoupling capacitor structures.

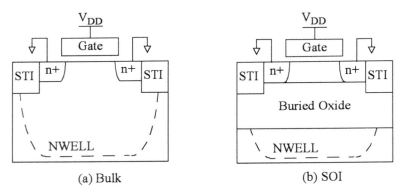

(a) Bulk   (b) SOI

FIGURE 2.10 Cross-section of a decoupling capacitor in bulk (a) and SOI (b).

One problem with both of these structures is that a defect in the thin oxide will cause a V_{DD} to GND short and draw current. Common practice places a control FET in series between the source/drain node and GND. Therefore, if a defect does occur in the thin oxide, the control FET can be turned off and the current from V_{DD} to GND is eliminated. However, turning off the device removes the decoupling capacitance as well.

2.8 Summary

To make use of SOI, one needs to have available all standard active and passive device structures. This begins as soon as the SOI wafer is created. Three common techniques are used; SIMOX, bonded and Smart Cut. All three are capable of producing thin silicon regions on top of a buried oxide.

Once the wafer has been produced, the devices are needed. The FET device can come in two flavors: partially depleted and fully depleted devices. The main difference is how charge is kept within the body of the FET. Partially depleted devices do not have as stringent a tolerance requirement in the thickness as the fully depleted device. The body of the FET is floating in SOI. If one wants to control the body, body contacts are available at the expense of increased area and increased capacitance.

Other supporting structures such as diodes and resistors are feasible in SOI, but the basic structure is changed. Diodes require a poly gate to form the p-n junction and will have a higher series resistance. Buried resistors and OP resistors are unchanged, but NWELL resistors do not exist. Decoupling capacitors maintain the same layout, but will have more series resistance to the source and drain. SOI has removed the inherent decoupling capacitance that was present at the junction between the NWELL and the substrate.

With device structures in place, let's now consider the electrical properties of these devices.

REFERENCES

[2.1]   ADVANTOX Spec sheet.

[2.2]   P. F. Lu, et al., "Floating Body Effects in Partially-Depleted SOI CMOS Circuits", *1996 International Symposium on Low Power Electronics and Design, Digest of Technical Papers*, 1996, p.139.

[2.3]   Neal Kistler and Jason Woo, "Scaling Behavior of Sub-Micron MOSFETs on Fully Depleted SOI", Solid-State Electronics, Vol. 39, No. 4, 1996, pp. 445-455.

[2.4]   L. Wei, et al., "Double Gate Dynamic Threshold Voltage (DGDT) SOI MOSFETs for Low Power High Performance Designs, *1997 IEEE International SOI Conference Proceedings*, 1997, p. 82.

[2.5]   S. Maeda, et al., "Substrate Bias Effect and Source-Drain Breakdown Characteristics on Body-Tied Short-Channel SOI MOSFET's," *IEEE Trans. Electron Devices*, Vol. 46, No. 1, January 1999, pp. 151-158.

[2.6]   D. Munteanu, et al., "Generation-Recombination Transient Effects in Partially Depleted SOI Transistors: Systematic Experiments and Simulations," *IEEE Trans. Electron Devices*, Vol. 45, No. 8, Aug. 1998, pp. 1678-1683.

[2.7]   A. J. Auerton-Herve, "SOI: Material to Systems", *Proceedings of the 1996 IEDM*, 1996, pp.3-10.

CHAPTER 3   *SOI Device Electrical Properties*

3.1 Introduction

So many currents in such a little puddle![1]

As described in the previous chapter, the SOI device is essentially identical to the bulk MOSFET well known to the industry, with the addition of an insulating layer under a thin active silicon region. This one, relatively simple alteration has a profound effect on every mode of the device's operation. At first glance, the behavior of the resulting structure is simple, and similar to the bulk MOSFET. A closer look reveals the complexity that this change creates. Given the description of the SOI structure in Chapter 2, lets examine the electrical responses it causes.

Central to the study of the SOI device are the multiple influences on the voltage of its electrically-isolated body, and the charge it contains. Given an understanding of these effects, it is possible to model and anticipate the behavior of the device under an array

---

1. From "National Velvet" with Elizabeth Taylor, one of Kerry's daughters' favorite movies.

of different operating scenarios. Do not despair, we'll carefully go through each later! For now, lets first establish their origin.

> *Coupling Capacitance from the drain or gate of a transistor to its body asserts profound SOI device performance effects.*
>
> *Source and drain junction leakage is the predominant passive transport mechanism of charge into and out of the body of the SOI MOSFET.*
>
> *Active diode action is the predominant active transport mechanism for instantaneously moving charge out of the body of the SOI MOSFET.*
>
> *At high voltages, impact ionization creates additional body charge.*
>
> *With continued scaling, gate oxide tunneling, and charge multiplication will also potentially alter body potential.*

We begin with a review of mechanisms which are key to floating body behavior.

3.2 SOI MOSFET's Junction Diode

The characteristics of the junction diode formed by the transition from a body or sub-strate to a counter-doped source or drain implanted region are central to the behavior of the MOSFET, and have even more significance in the PD-SOI MOSFET. Specifi-cally, *diode action, junction capacitance* and *junction leakage* strongly influence SOI performance.

At the heart of the PD-SOI MOSFET are two junction diodes formed by the counter-doped source and drain regions, identical in concept to those in the bulk device. The-body-source diode is often weakly forward biased; and the drain-body diode is usually reverse-biased. These diodes display classic diode behavior, as shown in Figure 3.1. Different fabricators use different doping profiles. The curves shown in Figure 3.1 is not specific to any particular company.

The capacitance between the source or drain and the body of the MOSFET is a strong function of their difference in potential. Referring to Figure 3.2, when the potential on the drain of an NFET is high and the body is low, the space-charge region surrounding the junction grows, as a larger counterdoped P area in the body must be "raided" for its electrons in order to satisfy charge neutrality across the junction. This space charge region acts as the dielectric in the capacitor formed by the drain and body. The dielec-tric, incidentally is silicon, with dielectric constant, ε_r, of approximately 12. At high potentials, the two plates are far apart and the capacitance is low. Conversely, at low differences in potential between drain or source and body, the resulting space-charge

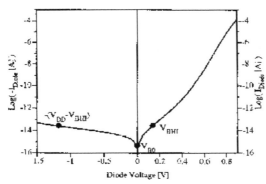

FIGURE 3.1 Classic diode I-V relationship for reverse (left) and forward (right) bias [3.1]

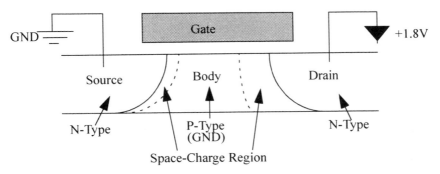

FIGURE 3.2 Space-charge region surrounding junction diode in NFET

region is thin, and the capacitance high. Figure 3.2 shows the relation between potential and capacitance for a NFET with its body at GND potential. Junction capacitance will be shown to be an important influence on SOI MOSFET behavior, as it has the ability to couple changes in gate, source, or drain potential into the body's voltage instantaneously.

Junction leakage refers to the low current passing through a diode even though it may be strongly reverse-biased. Junction leakage arises from three major mechanisms:

Electron/Hole recombination in the space charge region

Defects/impurities in the space charge region which disrupt the diode doping gradient from n-type to p-type.

HIgh-energy carriers which exceed the diode's electronic barrier height.

Junction leakage is sustained in bulk technologies by the supply rail connection to the substrate or N-well on one terminal, and the source or drain node driven by preceding circuitry on a second terminal. In the PD-SOI device's isolated floating body, junction leakage directly affects body potential and hence performance, over an extended period. Junction current also has an exponential dependence on voltage and temperature, as represented simplistically in the classic diode current equation,

$$I = I_0 (e^{qV/kT} - 1)$$ (3.1)

where I_0 is the diode's generation current with 0 volts across its junction, V is the potential across the junction, k is Boltzmann's constant, and T is temperature in degrees Kelvin.

3.3 Impact Ionization

Impact ionization of silicon lattice atoms at the pinched-off channel end of the transistor has been a device design concern in bulk CMOS. The problem it caused, however, was device wear-out. Performance was impacted gradually over the product's lifetime. In SOI, impact ionization of surface atoms continue to cause long-term wear out, but also have a more immediate effect on the device's performance.

Majority carriers (electrons in NFET devices, holes in PFET devices) have the ability to induce damage in the host FET structure when they are excited by high electric fields. The damage accumulates over the life of the product, raising the threshold voltage and reducing drain current in the NFET, and decreasing threshold voltage and increasing drain current in the PFET. This degradation affects performance, and can eventually cause the part to fail when the built-in timing margin of that path is consumed.

There are three modes of hot-carrier-induced degradation: conducting hot-carrier, nonconducting hot-carrier, and substrate hot-carrier. In SOI, we are predominantly concerned with the first two.

Conducting Hot Carrier Degradation occurs as the device is turned on. With increasing gate drive, the inversion region forms across the device channel. For gate drives below a critical voltage, the inversion layer is "pinched-off" at some potential, V_P, less than V_{DRAIN}. At pinch-off, the inversion region barely reaches the drain. The electric field across the excessively short uninverted remainder of the channel is then extremely high and essentially 'heats' the electrons to an effective high temperature, resulting in a distribution of kinetic energies with a high-energy tail. This scenario is illustrated in Figure 3.3.

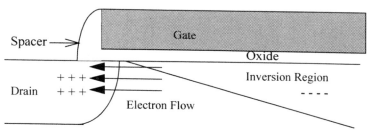

FIGURE 3.3 NFET at pinch-off, and resulting electric fields

While most electrons are collected by the drain, in the case of NFETs, a portion of the carrier energy distribution which is in excess of the gate insulator potential barrier height will be injected into the insulator, causing the familiar hot electron or "Hot E" wear out mechanism, described in more detail in other texts [3.2]. In addition to the damage at the silicon interface, impact ionization of silicon lattice atoms leaves behind positive charge, which accumulates in the isolated body of the device, and electrons, collected at the drain or gate regions. The high energy tail of the majority carrier energy distribution has the capability of ionizing atoms at the pinched-off (drain) end of the channel. Figure 3.4 schematically illustrates this process.

Non-conducting hot carrier degradation is ongoing device degradation which occurs when potential of magnitude V_{DD} is across the device (i.e. drain at voltage V_{DD} and source grounded for an NFET) but the gate voltage is less than V_T. Caused by the same mechanisms as conducting hot carrier degradation, a portion of the inversion carriers present in subthreshold or punchthrough current has sufficient energy to produce damage in the FET gate insulator oxide. The effect is most often seen in shorter channel devices, with low threshold voltages and high lateral electric fields. The same distribution of majority carrier energies putting negative charge in the gate oxide or into the drain also leaves behind positive charge which accumulates in the electrically isolated body of the PD-SOI device.

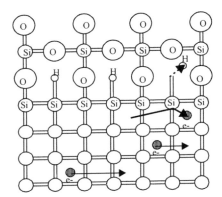

FIGURE 3.4 Representation of Si-SiO$_2$ Interface and Hot Carrier Event

Substrate Hot Electron Effects, which were caused in bulk silicon technology by the field from a charged gate to a grounded substrate, are absent for the most part in SOI, as the body quickly floats up in potential to essentially collapse this electric field.

With this background, lets now examine why a result of the first two mechanisms, **charge accumulation in the body region**, is significant in SOI.

3.4 Floating Body Effects

3.4.1 History Effect and Threshold Voltage Variability

The most prominent electrical property of the PD-SOI device is the *History Effect.* Simply stated, the I-V characteristics of the MOSFET built in PD-SOI are no longer constant, but dependent on the amount of charge contained in the body of the device at any given time. The charge content of the body, and the distribution of that charge caused by gate, source, and drain potentials determines the behavior of the device. Charge in the body is directly related to the potential of the body. The dependence of MOSFET threshold voltage on substrate bias is well known, and is displayed in Figure 3.5. Conceptually, body bias's effect on threshold voltage may be explained by how strongly this potential reverse-biases the junctions, which must be overcome by

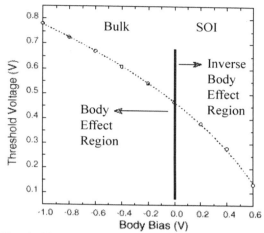

FIGURE 3.5 Threshold voltage dependence on substrate bias

gate drive. The magnitude of charge contained in the body is dependent on a number of factors which include:

Previous state of transistor

Schematic position of transistor (possible source, drain voltage ranges)

Slew rate of input, and load capacitance

Channel length and processing corner

Operating supply voltage

Junction temperature

Operating frequency and specific switch factor

A few basic concepts help explain the behavior of the body. Most importantly, three conduits convey charge into the body, and only two conduits conduct charge out of the body. Charge paths *into the body* are usually slow, on the order of magnitude of milliseconds to swing the body from rail to rail. Charge paths *out of the body* can be fast, essentially at the same rate as the device inversion layer formation.

Referring to Figure 3.6 below, charge slowly makes its way into the body via two means:

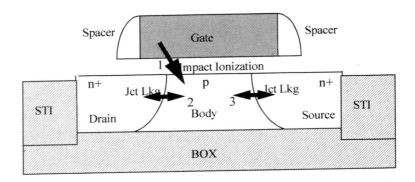

FIGURE 3.6 Charge paths in and out of PD-SOI NFET's body

1. Impact Ionization creates electron-hole pairs, some of which do not recombine (see Section 3.3). The electron is bound by the field created by the gate, but the hole is liberated into the body as positive charge. The holes will slowly accumulate, using the path shown by vector 1.

   Rule of thumb: Impact ionization has a substantial effect on body charge and body voltage at elevated V_{DD}, appearing as high V_{DS}. At nominal operating voltage, impact ionization asserts relatively minor effects on body charge or body voltage.

2. Junction leakage across the drain-body or source-body N/P diode slowly introduces additional charge into the body, illustrated by vectors 2 and 3. Over extended intervals, the amount of charge transferred can be substantial.

Charge exits the body via two mechanisms.

1. As the body slowly rises in potential via junction leakage or ionization charge accumulation, the source-body N/P diode and in some cases the drain-body N/P diode slowly become forward biased, and pass charge. The amount of charge passed through vectors 2 and 3 eventually levels off when a balance is achieved. This mechanism varies in its time to convergence, ranging in delay from tens to hundreds of microseconds.

2. A rising gate or drain will capacitively couple the body of the FET higher also. As the body rises at some fraction of the slew rate, the source-body N/P diode and in some cases the drain-body N/P diode will forward bias and quickly spill charge. (more on this later!). The typical path out of the body via this mechanism is vector 3.

3.4.2 The Body Potential Range

The range of possible body potentials for a given device varies with the given spectrum of voltages which the source and drain may assume, arising from the device's position in the circuit. An NFET with its source tied to GND and its drain reaching V_{DD} may never achieve a body potential greater than 0.5 V, while an NFET with both source and drain capable of sitting at V_{DD} may in fact have body potentials as high as V_{DD}. Hence it follows that the potential of a device's body at any given moment is the product of the device's recent use as well as its schematic position. Figure 3.7 provides qualitative insight into the equilibrium charge in the body of a floating-body MOSFET with its source tied to GND, and with its gate and drain as shown on the X and Y axes. Figure 3.8 provides the same equilibrium surface, but for a transistor with its source tied to the supply rail. As we will learn in the next chapter, the performance

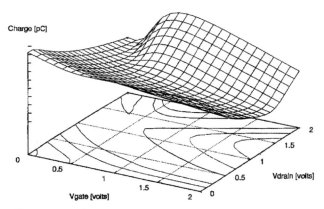

FIGURE 3.7 Charge contained in body, as a function of V_{DRAIN} and V_{GATE}, with source tied to GND.

and behavior of a given circuit is critically dependent on the body potential of its transistors immediately preceeding an input transition. As Figure 3.7 and Figure 3.8 show us, there are a number of permutations of gate, source, and drain voltages which will establish similar body voltages. Nonetheless, the device seeking equilibrium has the ability to move into dramatically different corners.

3.4.3 The Body-Charging / Device Wearout Connection

As discussed above, *impact ionization* provides a means of charging the body of a device. As shown in Figure 3.9, it has been observed that peak *conducting* impact ion-

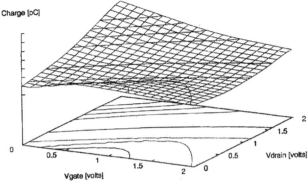

FIGURE 3.8 Charge contained in body, as a function of VDRAIN and VGATE, with source tied to V_{DD}.

ization occurs when the gate voltage is approximately $V_{DD}/2$. It follows then, that

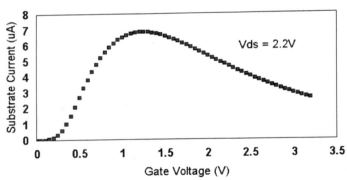

FIGURE 3.9 Impact Ionization as a function of gate voltage.

factors which prolong the device operation through this region will exaggerate the body potential and threshold voltage variability. Slow input slew rates, and/or heavy output load capacitances which slow lateral field collapse both cause body charging and increased history effects.

Rule of thumb: For NFETs electron hole pair generation is worst when $V_{GATE}=V_{DD}/2$ and $V_{DS}=V_{DD}$. Slow slew rates through this operating point, or high transistor switch factors cause larger accumulated degradation and charge generation.

Process/voltage/temperature (PVT) all have first order effects on these mechanisms which can vary body charge in the device's conducting mode.

Process variability alters the ratio of forward and reverse diode leakages, which will establish new balanced voltages. Shorter channels will also produce more impact ionization, resulting in more history effect. Conducting Hot Electron generation also simultaneously presents as a degradation in device current. This degradation's dependence on channel length, and hence the electron-hole pair generation for a typical production CMOS technology is shown in Figure 3.10 below. Shorter channels also produce bodies with less total volume. Smaller bod-

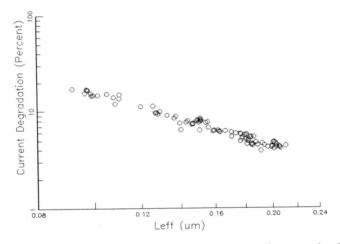

FIGURE 3.10 Device drain current degradation resulting from conducting hot-electron generation, in percent, as a function of channel length. SOI and bulk transistor produce similar effects for comparable devices. Note that degradation saturates for very short channels.

ies contain less charge, and the decreased volume reduces the time necessary to achieve large excursions in body potential.

Voltage of the supply affects junction leakage, and will affect the body potential. Of importance is not only the magnitude of forward and reverse leakage currents, but changes in the ratio of forward bias current to reverse bias current.

Temperature strongly affects junction leakage and device threshold voltage, as well. Lower threshold voltage at higher temperatures increases the portion of the electron energy distribution capable of ionizing silicon lattice points. This again affects the potential where the current into the body is balanced with the current out of the body. Temperature also affects the leakages of the junctions themselves, directly affecting body charge content.

The frequency and switch factor of a circuit's operation also have first order effects. PD-SOI MOSFET bodies achieve their potentials by operating in one of 3 modes:

Equilibrium State, when the device has sat with fixed potentials on gate, drain, source for an extended period, allowing the body to stabilize;

Steady State, when the device has been operated at a fixed frequency with stable drain and source voltages; and

Dynamic state, when gate or drain/source coupling to the body draws the body to a dynamically established potential.

As the device moves from one of these states to another, the history effects which dominate that operating mode will also emerge. If a given amount of charge is transferred into or out of the MOSFET's body on each stage transition in which that transistor participates, then as frequency or switch factor increases, the device will exhibit more variability in a fixed amount of time.

Although our interest in hot electrons above is motivated by its influence on body charge, classic device wearout due to conducting hot electron effect clearly occurs just as readily in SOI as it did in bulk CMOS. Figure 3.11 below displays original and stressed (degraded) current for a typical NFET. A device, such as a short channel N-

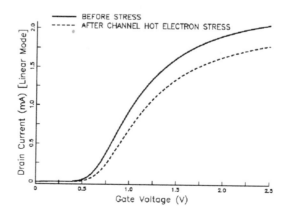

FIGURE 3.11 NFET channel hot electron wearout

MOSFET, which quickly and repeatedly achieves its maximum equilibrium body potential may suffer substantially more non-conducting hot-electron damage than in the bulk case or more than other devices on the SOI chip, due to increased leakage in its off-state induced by substrate bias.

3.4.4 The "Kinks[2]"

It was discussed above that as a given PD-SOI MOSFET transitions from accumulation into inversion and saturation, it moves through an interval of gate drive in the *conducting mode* where impact ionization peaks, generally at $V_{DD}/2$. The injection of positive charge into the body has a noticeable effect on the dynamic behavior of the device. Because a large sudden increase of positive charge will reduce threshold voltage, a *kink*, or increase in I_{DS} may be observed when the gate voltage reaches approximately $V_{DD}/2$, as shown in Figure 3.12. This change in slope, observed at normal

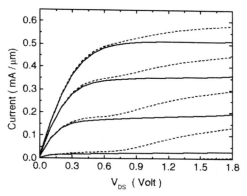

FIGURE 3.12 The First Kink on PD-SOI MOSFET I-V Curve.

operating voltages, is often referred to as the "First Kink."

A second kink, not nearly as noticeable as the first kink occurs after the first kink. As the device current increases, the body-to-source diode can eventually forward-bias, enabling the flow of bipolar current in the structure. This bipolar device, in parallel

---

2. No relation to Lo-Lo-Lo-Lo-Lola.

---

with the intended MOSFET, tends to augment the MOSFET. For most practical applications, this second kink is usually overlooked.

A third kink, of sorts, may be observed when the device is operated at elevated voltages, as practiced during reliability stress testing. As the supply voltage increases, the kink just described first appears at lower and lower gate voltages; at a sufficiently high V_{DD}, impact ionization in the device's *non-conducting mode,* with the gate voltage at GND, causes positive charge to accumulate in the NFET body, depressing threshold voltage before the gate is even turned on. The second kink is a concern when stressing parts at elevated voltage and temperature, when functionality is still required. The naturally lower threshold voltages caused by elevated stress temperatures exaggerates this impact ionization, further threatening circuit functionality.

The kink, as discussed above, occurs because of impact ionization which charges the body of the device and raises body bias. It follows, then, that the more time the gate potential spends in the device region of operation which maximizes impact ionization, the more noticeable the kink will be. Another way of saying this is that the "amount" of kink would be expected to vary with gate switching frequency. This is, in fact, the case and is exhibited in Figure 3.13 [3.5].

1. Drain current measured at DC conditions exhibits a pronounced kink effect, which becomes less apparent on pulse I-V plots taken at various frequencies because of the mechanism just described.

2. The kink's strong dependence on gate bias can also be seen. At higher gate voltages, the kink disappears entirely. This is expected, as high overdrive forms a solid inversion region. This avoids the device scenario which enhances impact ionization at the drain end of the channel.

3. Because the kink relies on elevating majority carriers past their ionization potential, the minimum V_{DS} where the slope of the I-V curve "breaks" upward remains relatively constant, at roughly $V_{DD}/2$.

4. Finally, the reader may have noticed in Figure 3.13 that the drain current from pulsed devices exceeds that of the device measured in the DC case for all but the lowest gate bias. This difference is a result of the mode each device was evaluated in. For the DC curves, the body is always at *equilibrium* in any of the points on the plots; for positive gate voltages, the channel is conducting which reduces the junction potential and with it junction leakage. In the pulsed device, however, its body is at a *steady state.* In this case, with the device off for a portion of the cycle, junction leakage will charge the body higher during the off state, resulting in lower threshold voltages and more overdrive when the gate transition finally comes

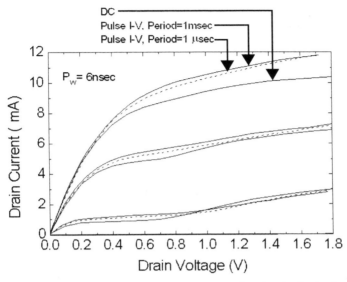

FIGURE 3.13 DC and Pulse I-V Curves for a 0.25μ floating body device [3.5]

along. Steady state body potentials usually are higher than equilibrium body potentials, but again vary with the device design point. See your dealer for more details.

3.4.5 Elevated DIBL

Drain-Induced Barrier Lowering (DIBL) is another well known MOSFET mechanism [3.3] which changes with the introduction of the floating body. For long-channel devices, above a relatively low V_{DS} (approximately $2kT/q$) with 0 V gate bias, the effect of the lateral drain-to-source electric field on drain-to-source device current saturates, and device subthreshold leakage becomes relatively independent of V_{DS}. For short-channel devices, however, DIBL results in an exponentially decreasing barrier near the source, and thus increases subthreshold current. Another way of thinking of this as drain voltage increases, the drain-to body junction diode expands its space-charge depletion region. The remaining channel is shorter, resulting in short-channel-effect-like barrier reduction.

In a PD-SOI device with a floating body, DIBL becomes artificially high, and to some extent loses its meaning. With the floating body potential able to rise as drain charge leaks through the drain-body junction diode, the threshold voltage of the device lowers, complementing the natural barrier lowering of normal CMOS at shorter channel lengths. As the body voltage rises, the channel length actually remains long because the space charge region surrounding the high-voltage drain shrinks. This subdues short channel effect threshold voltage reduction. Nonetheless, the body voltage has

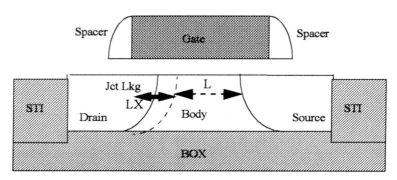

FIGURE 3.14 Competing effects of drain depletion and threshold dependence on body potential. Channel length L shrinks with higher drain potential, reducing device threshold voltage. Leakage LX raises body potential which lengthens L but depresses threshold voltage.

the larger influence on threshold voltage, and so the device appears to have large DIBL. Depending on process, an SOI device with its body contact connected to a stable potential will revert to the bulk-CMOS DIBL values.

3.4.6 Reduced Body Effect / Source Follower Action

An important electrical property which accounts for some of SOI's performance advantage is the reduced body effect exhibited by MOSFETs arranged in series.

To review CMOS Body Effect, refer to Figure 3.15. Due to voltage drop across device T1, potential of device T2's source is elevated above GND while current is passed. T2, in absolute terms, sees negative substrate voltage with respect to its source (Voltage V_{SSx2} is negative), making T2's channel region harder to invert. Higher source-to-substrate bias V_{SSx2} increases threshold voltage V_{T2}, and the circuit loses performance.

PD-SOI transistors without body ties exhibit "Inverse Source Follower Effect," in which the body actually floats up in potential a diode voltage drop above the source

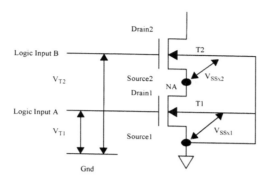

FIGURE 3.15 CMOS Body Effect. In Bulk CMOS, all NFET devices have their bodies tied to chip GND. In PD SOI this connection is optional

potential. With the transistor body typically at a higher potential than GND, voltage VSSx2 is reduced, and stacked device performance is improved.

> **Rule of thumb:** In PD-SOI, an additional evaluate device may be added to a stack without exceeding the delay of its Bulk-CMOS predecessor.

Unlike the intrinsic performance improvements contributed by mechanisms described previously, body effect improvement may not be fully exploited on many designs due to a number of reasons:

1. Circuit designers have been conditioned from their bulk CMOS experience to avoid the threshold voltage drop associated with stacked devices. This is a hard habit to break, although it will allow considerably more logic content per stage.

2. A stage with multiple stacked circuits will always be slower than a stage with one or two devices in series. In few cases do all the inputs in a multi-device evaluate stack arrive at the same time: usually the last arriving signal reaches the given circuit well after most of the earlier signals have stabilized. In that scenario it is usually faster to "pre-evaluate" the earlier signals, and pipe the output to a smaller stage which integrates the intermediate result with the latest arriving signal. The result is less total delay.

3. SOI designs frequently are map-overs from Bulk CMOS, done to quickly secure extra performance. Because taller NAND stacks are not used in bulk CMOS, they naturally do not appear in the SOI version.

3.4.7 Reduced Short Channel Effect

Another legacy characteristic from Bulk-CMOS is the threshold voltage dependence on substrate bias. The short channel effect refers to threshold voltage reduction with channel length for very short channel devices, and is shown generically for a bulk device in Figure 3.15.

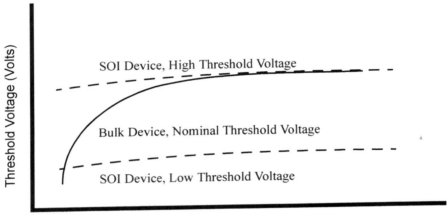

Channel Length (microns)

FIGURE 3.15 CMOS Short Channel Effect, showing roll-off of threshold voltage with L_{eff} for bulk and SOI devices

For these devices on the short end of the channel length distribution, the magnitude of short channel threshold voltage roll-off has been observed to decrease with elevated substrate bias, as was shown in Figure 3.5 on page 35. It follows, then, that as the body voltage in our floating body, PD-SOI device increases in voltage, the short channel threshold voltage roll-off decreases or disappears. Figure 3.15 superimposes the saturation threshold voltage dependence on channel length for the SOI MOSFET, with the body left floating, for high and low threshold options. With less threshold voltage roll-off with channel length, less variability is also observed in SOI performance across the process window of L_{eff}. In Bulk CMOS, both the on-resistance and the threshold voltage of varying channel lengths contributed to wider performance variability in a given circuit. In SOI, channel length variability to a first order affects only the device's on-resistance, tightening the resulting circuit's range of possible delays.

When drafting process settings which define device thresholds, the process engineer must select a value which, when rolled off to its minimum voltage at shortest channels, presents acceptable leakage and noise immunity characteristics. The absent short channel roll-off shown in Figure 3.15 allows the process developer to define a lower average threshold, which enhances overall performance at long channels without exposing design parameters such as off current or noise margin at short channel lengths.

3.4.8 Bipolar Device Action

Another noted "feature" of the partially depleted SOI device is the presence of a parasitic bipolar junction transistor. In fact, the bipolar transistor has always been present in CMOS. In bulk CMOS, the snap-back and latch-up concerns arise from this duality. An SOI device with a floating body merely makes the presence of the bipolar parasitics substantially more prominent.

Figure 3.16 displays the structure of a common bipolar junction transistor (BJT). In

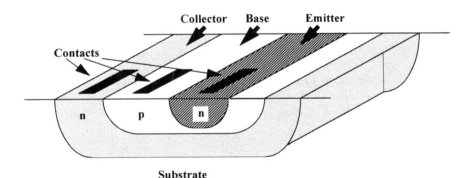

FIGURE 3.16 Common Bipolar Junction Transistor

an NPN BJT, electrons flow by diffusion across the base region from emitter to collector under the presence of a potential difference from collector to base. In Figure 3.6, "Charge paths in and out of PD-SOI NFET's body," on page 36, it can be noted that the MOSFET structure also forms a bipolar structure, where the drain acts as the BJT collector, the body as the BJT base, and the source as the BJT emitter. What is unique to SOI is that the floating body of the SOI device can rise sufficiently high in voltage with respect to the source or drain to forward bias the respective junction diodes, permitting bipolar gain.

Referring to Figure 3.17, if the body of the transistor has been charged high (due to

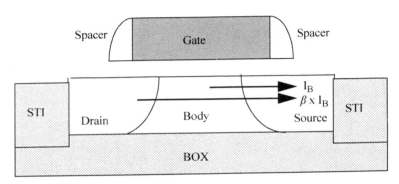

FIGURE 3.17 Bipolar device action in SOI

past use) and the source of the device is suddenly drawn low during a switch, the body-source junction diode can become instantaneously forward biased, and will conduct. Then current I_B momentarily flows from body(base) to source (emitter) until the charge in the body is sufficiently discharged. Simultaneously, bipolar action causes a current of $\beta \times I_B$ to briefly pass from drain (collector) to source (emitter). Device design attempts to keep the value of β low, but depending on the value of V_{DD}, this bipolar gain can be above 1. If the charge in the drain is not actively held high (as in some dynamic logic styles), then charge lost to bipolar action can be substantial. We will spend the majority of Chapter 5 exploring this exposure in more detail.

The peak and duration of the bipolar current is a predominant concern to designers. If the bipolar current is too large, the loss of charge on a given node can create logical fails in dynamic and passgate circuits. Consider the NFET whose gate is tied to GND and the drain is tied to V_{DD}. If the source of this device is attached to pulse generator with a constant frequency, the bipolar device will be turned on and off as the source is pulled low. Figure 3.18 shows the peak bipolar current as a function of frequency for two different channel lengths. Here the bipolar current is largest when the period of the input frequency is longest. This allows the body to charge to its highest potential before the source is transitioned toward GND, creating the most charge in the body and resulting in the largest peak bipolar current. The amount of time needed for this device to reach DC equilibrium is about one to ten milliseconds. If the source is toggled at a much shorter period, the potential on the body fails to approach V_{DD} and the bipolar current is smaller.

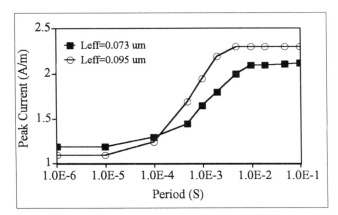

FIGURE 3.18 Peak bipolar current as a function of input pulse period for two different channel lengths.

As the channel length is shortened, there are two competing factors that determine the peak bipolar current. At a shorter length, the body can hold less charge and thus the peak current tends to be reduced. However, as the length is shortened, the equivalent base of the bipolar device is shortened, thus increasing the gain of the bipolar device. For this FET, the shorter device length resulted in a lower bipolar current at long periods and higher bipolar current at high frequency, or long periods.

Since bipolar current can make substantial impacts on circuit response, one should try to limit its magnitude. One means of achieving this might be to increase the reverse saturation current of the junction diodes, to achieve body potential stability more quickly. Outside of the device design, application conditions can help to reduce the circuit sensitivity to the bipolar current. Figure 3.19 shows the peak bipolar current as a function of voltage for two different temperatures. As the voltage and the temperature is reduced, the magnitude of the bipolar current is reduced. This assists one with the use of voltage and temperature to ascertain if a circuit sensitivity is due to bipolar current or some other circuit phenomenon.

Note that this discussion has only been for the NFET. The PFET also has a parasitic bipolar device which enables a parasitic response in a similar manner to the NFET. Due to the doping that occurs creating the NWELL, the gain of the parasitic bipolar device for a PFET is less than the gain of the parasitic bipolar device for an NFET. Besides exhibiting a smaller response, most evaluate trees are built with NFETs for superior performance; stacking these NFETs is what enables the bipolar effect. As will be discussed in the next several chapters, the bipolar current within a PFET does

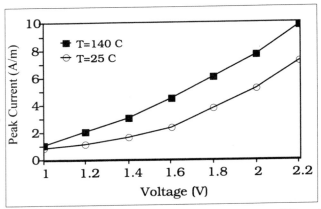

FIGURE 3.19 Simulated Bipolar current curves vs. V_{DD} at two different temperatures, for a typical process.

not come into consideration due to the nature of passgates circuits, SRAM structures and dynamic circuits.

As one elevates the drain to source voltage high enough, the bipolar device will eventually breakdown. In bipolar breakdown avalanche currents overwhelm the barriers established by the source and drain implants, quickly creating exponentially large currents which pass from drain(collector) across the body(base) to the source(emitter). Figure 3.20 displays an I-V curve for an NFET. Notice that the bipolar breakdown is above 2.5V. This is well above functional operating supply voltages, but is within the window of present voltage stress conditions such as burn-in[3]. Under burn-in conditions, the total current from the bipolar devices breaking down is prohibitive and can create conditions where the chip tries to sink several hundred amps. This causes one to reduce the stress voltage to 2.5V or below. As the supply voltage continues to scale down below 1.5V, the breakdown of the bipolar device will move farther away from the operational supply and stress voltage ranges, providing some relief and freedom in determining reliability stress conditions.

Breakdown voltage is also a function of channel length. At longer channel lengths, the base of the parasitic bipolar device is also longer, requiring a higher effective voltage to recreate the electric filed necessary to induce breakdown. Figure 3.21 shows an experimental SOI NFET's breakdown voltage dependence on channel length.

---

3. Conventional Burn-In technique prescribes device operation at 1.5X V_{DD} and 140 degrees C for an extended period of time.

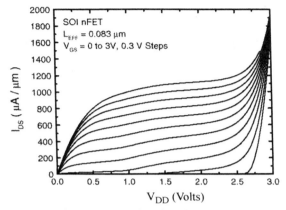

FIGURE 3.20 I-V curve for an NFET showing the onset of bipolar breakdown above 2.5V.

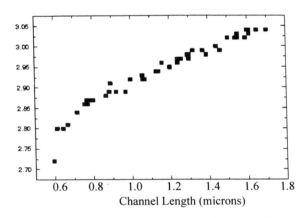

FIGURE 3.21 Bipolar breakdown voltage as a function of channel length.

3.4.9 Lower Temperature Sensitivity

The response of the PD-SOI device to temperature is **highly dependent** on the device design, and so it is only with trepidation that generic observations on temperature response are offered here. Two compensating PD-SOI characteristics, however, contribute to reduced MOSFET sensitivity to temperature.

1. As described on page 50, body-source and body-drain junction leakage is intentionally increased to subdue floating body effects in SOI. In addition to this leakage, increased junction temperature naturally increases this junction leakage further. Higher junction leakage reduces the equilibrium body voltage in cases when the source is lower in voltage than the drain. This lower body potential results in higher threshold voltages, countering the conventional thermal subthreshold voltage reduction which is endemic to bulk CMOS. The end result is more uniformity in off-currents across temperature.

2. Device designers have the ability to reduce nominal threshold voltages in the PD-SOI device, as floating body voltage equilibrium results in for approximately half the short channel effect and temperature sensitivity as bulk. An obvious opportunity to improve performance, then, is to reduce threshold voltages enough to consume the extra minimum threshold voltage margin formerly required in bulk CMOS. This minimum threshold voltage is determined by both reliability and functional limits [3.4].

This leads us to the original question: how does this response reduce temperature sensitivity? With default threshold voltages uniformly lower in the product, leakage currents are already elevated with respect to bulk, resulting in less sensitivity and a lower percentage change in leakage.

3.4.10 Backchannel Device

A final subtle electrical property of the SOI device is the presence of a backchannel device. THe PD-SOI BOX insulator, although very thick, can nonetheless act as a *gate oxide*, and the substrate as a surrogate gate. If sufficient potential exists in the substrate, an inversion channel may be formed at the bottom of the floating body. The resulting device is much longer in length and has little practical use. Deep backside doping to elevate the threshold voltage of this undesired device, and grounding of the underlying substrate are two effective means to incapacitate this parasitic device.

Rule of thumb: Backside threshold voltage implants should produce backside threshold voltages of at least 4X V_{DD}.

The authors can testify from direct experience that, unless well controlled, the back channel device can and will cause absolute havoc in a design!

Even with the backchannel device turned off, the same materials interface issues which plague the front gate can provide a leakage path across the back device. As previously presented in Figure 3.4 on page 34, a dangling bond shown filled by hydrogen provides a potential leakage path supporting currents from drain to source in the nanoampere range. Surface states, trap sites and interface impurities all provide an opportunity for leakage.

3.5 SOI MOSFET Modeling

No good can come of the SOI device design point if it cannot be modeled well and used by circuit designers. The new effects that we have mentioned need to be accurately represented by the device's simulation model, so that the circuit designer can accommodate SOI's idiosyncrasies and take full advantage of the device's performance. We will investigate the model under DC and AC conditions.

3.5.1 DC Discrete Element Representation

The most important part of a DC solution for a model of the FET is the leakage currents of the junctions and their effects upon the voltage of the body. Figure 3.22

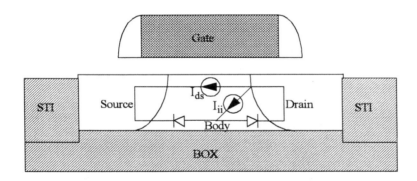

FIGURE 3.22 DC equivalent model for an PD-SOI NFET.

shows the DC equivalent model. Here the currents of interest are the leakage current, I_{ds}, the impact ionization current, I_{ii}, and the current through the two diodes. Depending upon the voltages of the source and drain, the body-source and drain-body diodes may each be reversed-biased or forward-biased. The body voltage will settle at a voltage between V_{DD} and GND in the equilibrium state. As we will see in the next section, this range can temporarily be exceeded by up to a diode drop V_d *above* V_{DD} or *below* GND in the AC case. This is a wide but fully realizable range.

3.5.2 AC Discrete Element Representation

A simple AC equivalent model is shown in Figure 3.23. A capacitive network exists

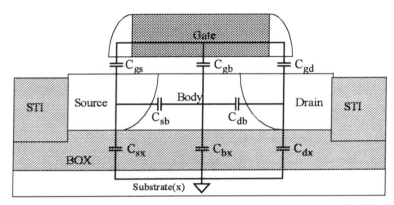

FIGURE 3.23 Simplified AC model for a PD-SOI NFET. The parasitic bipolar transistor is not shown in this model.

across the dielectrics of the thin oxide and the BOX as well as across the junctions that bound the body of the FET. With the creation of the floating body, the capacitance that leads to the body dynamically moves the body as the gate, source and drain are transitioned. This will be explored further in Chapter 4.

One important feature that this AC model doesn't show is the parasitic bipolar device that connects the source and drain through the body of the FET.

These two models are very useful when considering one's need to monitor body potential before, during and after the FET transition. One should understand the magnitude of the capacitances involved to best follow the voltage on the floating body.

3.6 Insulator-Related Effects

Up to now, the device behaviors we've examined have been associated with the electrical behavior of a MOSFET device with its body floating. In this section, we take a look at phenomena associated with the physical properties of the insulator creating this floating body. In general, the properties observed result from the isothermal, dielectric, and amorphous nature of silicon dioxide. What we'll conclude is that simultaneously, SiO_2 is both the best and the worst candidate for the job.

3.6.1 Self Heating

In bulk technologies, heat generated by charge transfer in the transistor was readily transferred out of the chip backside through the crystalline substrate. This transfer of heat was quick enough that local device transconductance changes due to self-heating was negligible. Heat was impeded from transfer up through the metallizations by the silicon dioxide used as the inter-layer dielectric (ILD). SiO_2 is a superb thermal insulator, limiting power transmission to 10.7 mW/cm degK. With 6 or more layers of interconnect, the stacked ILDs presented substantial thermal resistance. With the advent of reduced dielectric constant dielectrics, this value will drop even lower, to values as low as 2mW/cm °K.

In PD-SOI, silicon dioxide comprises the BOX layer. The PD-SOI transistor is now encased in a perfect little insulated region of its own, much like a cold drink in a picnic cooler, with SiO_2 ILD above, TEOS[4] to the left and right, and BOX underneath. As a result, the average junction temperature of SOI devices can be somewhat higher than an identical bulk device, reducing device transconductance. In the worst case, with a device on for 100% of a cycle conducting maximum current, perhaps 1 mW/micron may be dissipated by the device, causing a 60-100 degree C temperature rise above ambient.

> **Rule of thumb:** In the vast majority of cases, PD-SOI CMOS devices see changes in junction temperature caused by switching of 3 deg C or less.

In normal AC operating mode, where the device swings from cut-off to saturation, this temperature change has a negligible effect on performance and may be disregarded for all but the most heavily loaded nets with the highest switch factors, such as I/O drivers. In these circuits, substantial changes in the voltage and current slopes are evident with different loads [3.5]. The change in load drives different slew rates and heating amounts. This effect is displayed in Figure 3.24.

> **Rule of thumb:** Assume temperature increases of 15 degrees C for devices found in the final stages of I/O drivers.

When CMOS devices are biased to intentionally operate in their linear region, the temperature increases which occur during a given transition can cause the device to

---

4. TEOS (Tetra-Ethyl-Ortho-Silicate) is a commonly used, low quality silicon source from which an oxide is formed. Using low pressure chemical vapor deposition, it fills the isolation trenches etched into the silicon which separate adjacent transistors.

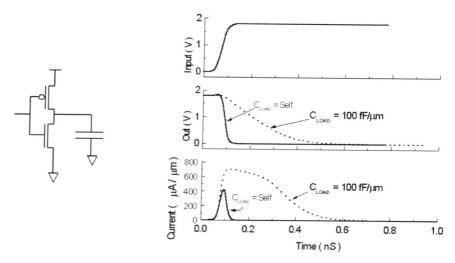

FIGURE 3.24 Self-heating effects on SOI device transconductance. Substantial changes in device behavior may be observed if the device stays on long enough to heat up appreciably.

depart substantially from its expected behavior. Unfortunately, little can be done to reduce this effect, other than to reduce currents. These thermal device effects especially complicate the design of common analog circuit elements such as phase-locked loops and delay lines found in microprocessor designs. SOI analog design concepts encountered in PLL design are discussed in more detail in the literature [3.6].

Devices which share SOI silicon wells also should be examined for potential differences in heat dissipation. Consider two devices, 1 and 2, which couple either Signal A or Signal B to GND, as shown in Figure 3.25. The devices have been laid out to use a continuous diffusion area. If device 1 has a low AC switch factor while device 2 has a high switch factor or prolonged DC current, the heat generated by device 2 is readily transferred through the crystalline silicon diffusion to the channel region of device 1, reducing its transconductance as well, even though it has little activity. To properly model this, one needs to thermally connect devices 1 and 2 together to share the same temperature as well as sharing the same source voltage. In contrast, device 3 is of much less consequence to devices 1 and 2, or vice-versa, as STI provides substantial thermal isolation.

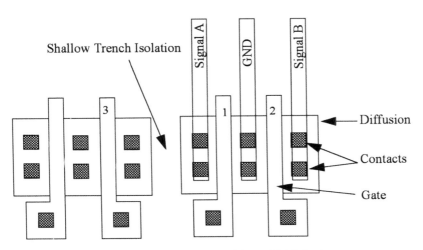

Shallow Trench Isolation

Signal A

GND

Signal B

Diffusion

Contacts

Gate

FIGURE 3.25 Thermal Isolation and Pairing of SOI devices

3.6.2 Diffusion Capacitance Reduction

As first discussed in Chapter 1, fully half of the performance advantage of SOI lies in the reduced junction capacitance it offers. Referring to Figure 3.17, "Bipolar device action in SOI," on page 48, the bottom area of the source and drain diffusions in the SOI device abut the buried silicon dioxide BOX layer. In a conventional bulk structure, this bottom area would abut the silicon substrate. Silicon dioxide has a dielectric constant (K) of approximately 4, while that of silicon is approximately 12. Hence the area component of the junction capacitance is dramatically reduced.

The perimeter capacitance, comprising the drain-body and source-body interface remains high. As described above, with the area capacitance now out of the picture, the source's and drain's capacitive divider influence on the body is now substantially stronger compared to the gate of the device.

3.6.3 Latch-Up Elimination

As mentioned in Section 3.4.8, parasitic bipolar devices are not new to CMOS. While the floating body SOI MOSFET introduces an especially hazardous bipolar mechanism, ironically it eliminates another.

Latch-up refers to the pair of parasitic bipolar devices formed by a PFET/NFET device pair in close proximity in bulk technologies. When perturbed by signal undershoot or overshoot, enough carriers are injected into the substrate to forward-bias one of the junctions, causing bipolar gain to be passed in that device. Resulting voltage drop during the current spike allows destructive feedback of the two bipolar devices to latch each other into the on-state. The latch will pass excessive currents until one of the supply interconnects fails due to joule heating. Figure 3.26 shows the implant profile of a bulk MOSFET. Figure 3.27 give a schematic representation of the parasitic bipolar transistors formed by the implants. The reader should note that these devices form a latch structure which, with sufficient gain, will latch in the "on" state, passing DC current which can damage the supporting interconnect.

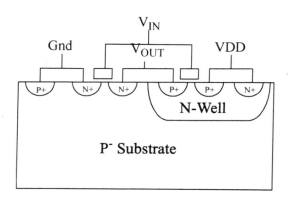

FIGURE 3.26 Bulk implant profile for the simple inverter [3.7]

SOI avoids latch-up by surrounding each MOSFET with dielectric. There is simply no path or structure which exists between NFETs and PFETS in a configuration which could produce bipolar currents during transitions. Referring again to Figure 3.27, each of the parasitic resistors and bipolar transistors are avoided by the presence of the BOX buried oxide layer.

CMOS Latch-up is a big deal in bulk technologies, and its elimination in SOI is a noteworthy feature. With the scaling of CMOS feature sizes, the pressure on increasing density, and higher implant doses, the propensity for latchup is real and threatens product reliability. PFET-to-NFET spacing can be made much tighter without the latchup concern.

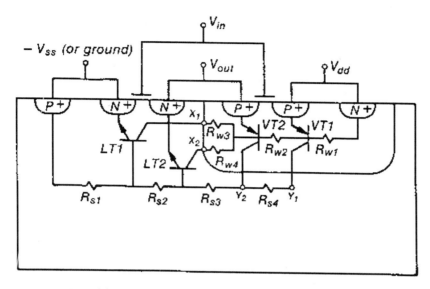

FIGURE 3.27 Parasitic bipolar devices arising from implant profile shown in Figure 3.26 [3.7]

3.6.4 SER Improvement

SOI technology has long been popular in extraterrestrial electronics applications due to its increased immunity to radiation-induced logic errors. In addition, even in earth-bound applications, the fail rate due to alpha and cosmic radiation has been steadily increasing as scaling reduces the capacitances on each node. As we will discuss in Section 6.7, "Soft Error Upsets" on page 141, radiation events cause less charge generation and collection in SOI, as the sensitive active region through which the alpha particle passes is limited to silicon above the BOX. Once the trajectory passes into the BOX layer or the underlying substrate, it can no longer effect the active silicon layer. The thickness of epitaxial silicon in the bulk device delivers much more charge. The particle creates ionized charge along its trajectory through the device, which is swept up by the reverse biased junctions.

3.7 Composite Responses

The circuit designer now has a device with a new personality to deal with. In addition to the first order effects described, a few design idiosyncrasies resulting from the composite SOI structure should be evaluated for their potential effects.

3.7.1 Device Leakage - 5 Mechanisms

To draw together a key point of this chapter, there are now 5 leakage transport mecha-

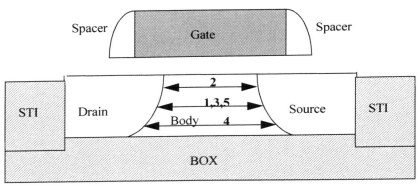

FIGURE 3.28 Five current transport mechanisms in PD-SOI

nisms in PD-SOI which must be monitored by the circuit designer as well as the technology process developer. These current paths, whose relative locations within the device are shown in Figure 3.28, comprise

1. Conventional punch-through current
   The use of device S/D extensions or LDDs is still essential in PD-SOI.

2. Conventional subthreshold leakage current, but at lower threshold voltages
   This is the predominant leakage contributor in both SOI and bulk technologies

3. Bipolar Current
   The parasitic bipolar junction transistor is in parallel with the FET and responds quickly.

4. Backchannel Device Current
   Robust PD SOI devices will include implants to elevate the backchannel threshold voltage.

5. Bipolar Breakdown Current
   Breakdown occurs at higher voltages, and is a challenge to Burn-In and lifetime acceleration stressing for defect discovery.

Again, the voltage necessary to establish each of these paths and the resulting current are specific to the process. In general, however, each of these mechanisms become more severe with elevated voltage, and must be planned for if the part is to be stressed.

3.7.2 Conservation of Charge

Keep your eye on the charge!

The dynamic redistribution of a fixed quantity of charge contained in the floating body of an SOI MOSFET is a fundamental concept which, if remembered, aids in the interpretation of its responses. Simply stated, charge must be conserved in the operation of the SOI MOSFET. PD-SOI has no special permission to violate Kirchoff's Law that the authors know about.

Because so much of the MOSFET's performance is associated with its capacitances between source, drain, channel, gate, and substrate, it follows then that changes in charge distribution affecting these capacitances will show up in performance. In Figure 3.29 below, a cross section is shown of a PD-SOI NFET in accumulation (1 in the diagram) and in inversion (2 in the diagram), displaying the change in distribution of minority carrier charge. Because the capacitance of the junction goes inversely with the width of the space-charge (depletion) region, this way of looking at the device helps to remind us of the impacts of redistribution of the charge in the body.

Lets consider a few simple examples. With the device in accumulation, charge in region 1 is quite close to the gate. As the gate rises, it easily couples the body high, and predominates in the influence on the body. But as the device turns on, charge redistributes into region 2, where the body's behavior is now more strongly influenced by changes in drain voltage.

One final note. Assessing charge redistribution, and considering the conservation of charge in the body of the device is facilitated by the AC view of the SOI MOSFET shown in Figure 3.23 on page 54. This by no means implies that this charge is an unchanging fixed amount. The reader is reminded that DC elements represented schematically in Figure 3.22 on page 53 are constantly modifying the total charge contained. But for comparing charge effects on the device in various modes, this approach is useful.

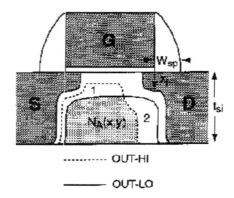

············ OUT-HI

———— OUT-LO

FIGURE 3.29 Body Charge Balance Distribution in the PD-SOI NFET [3.1]

3.7.3 Other Parasitic Active Elements

The presence of multiple adjacent dopings above an insulating layer gives rise to additional parasitic structures which can be formed when stacking MOSFETs in a NAND configuration. While they are not likely to cause problems in well-engineered devices, these oddities are pointed out for completeness.

The NAND structure shown in Figure 3.30a below places multiple MOSFETs in series using minimum uncontacted space between gates to minimize power, delay, and area. This minimum space reduces the volume of the intermediate source/drain diffusion NA in Figure 3.30b. In addition, the dose and energy of the implant received by this node may also be reduced by the "shadow" created by the gates to either side. Since the intermediate diffusion is not contacted, its potential is derived from the operation of the NAND itself. It follows, then, that any active parasitic elements gated by this node could provide a feedback response in this configuration. Two are described below.

1. *A parasitic JFET* may be formed when the intermediate source/drain diffusion is too shallow [3.8]. The body of device T2 acts as the drain and the body of device T1 acts as the source of this device. Elevated voltage on the intermediate diffusion NA causes the space-charge region surrounding the diffusion to expand down beyond the buried oxide layer, cutting off a conducting path between the bodies of T1 and T2. When NA goes to GND, however, the space charge region shrinks, and

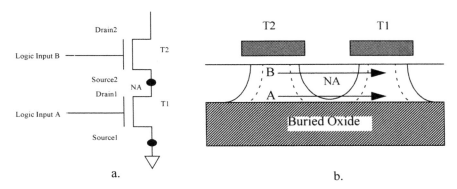

FIGURE 3.30 Parasitic structures unique to the SOI NAND configuration: The NAND NFET path (a); The SOI NAND cross-section (b).

may provide a path "A" in Figure 3.30b, albeit at high resistance, under NA. The resulting body voltages would not be at potentials anticipated by conventional modeling.

This parasitic element may also be modeled as an *unintended PFET*. The bodies of devices T1 and T2 can act as source and drain, and diffusion NA can serve as the body of this PFET. The gate electrode in this case is the substrate under the BOX. Because the gate capacitance to the channel through BOX is quite low compared to the inversion channel capacitance to the NA "substrate," it follows that this parasitic element has extreme "substrate sensitivity," agreeing with the sensitivity attributed above to NA as the JFET gate.

2. *A new parasitic bipolar* device may also be formed [3.8], its base being the intermediate diffusion D1. In the NFET pair shown, the resulting parasitic P-N-P bipolar is formed with the body of transistor T1 acting as a collector, the body of transistor T2 acting as an emitter, and the diffusion NA acting as the base. The resulting current path "B" is shown in Figure 3.30b. The size and relative dopings result in a fairly low-gain bipolar device, but again can result in body voltages not anticipated by the model. The reader is reminded that this is the second parasitic bipolar device potentially formed by the SOI structure; the first is the more prominent bipolar device which causes the source, body, and drain of the MOSFET to behave as emitter, base and collector.

3.8 Summary

The simple introduction of an insulating layer which isolates the body region of a conventional CMOS device from substrate bias gives rise to a number of new passive and active elements contained within the MOSFET structure. The capacitance and currents of the junction diodes formed between the source and drain implants and this isolated floating body, and ionization of silicon lattice atoms caused by electrons at elevated energy are responsible for the peculiar behavior unique to SOI devices.

SOI devices exist in one of three states; equilibrium state, steady state, and dynamic state. Each of these states induces different delays and behaviors in the many circuit topologies found within a given chip. The next chapters are, in fact, devoted to exploring the response of these different topologies to SOI floating body effects.

In addition to the behavior in the active silicon, the properties of the dielectrics used in SOI fabrication become even more important than they were in bulk silicon. Dielectrics assume a greater role in SOI. Silicon dioxide's lower dielectric constant is responsible for approximately half of SOI's advantage. Its inferior thermal conductivity, however, presents a new design constraint to the engineer.

Finally, a multiplicity of unexpected parasitic elements can be formed by the SOI structure. These parasitic elements include 2 different opportunities for bipolar devices, JFETs, and a second, longer-channel MOSFET.

The liabilities which come with SOI reaffirm the old adage that there is no "free lunch"; while SOI's performance improvements are attractive, they come with a price tag. Given the diminishing returns of scaling noted in Chapter 1, this price tag is looking more and more like a bargain!

REFERENCES

[3.1]   A. Wei, et al., "Design Methodology for Minimizing Hysteretic Vt Variation in Partially Depleted SOI CMOS," *Proceedings of 1997 IEEE International Electron Devices Meeting*, pp.411-414.

[3.2]   C. Hu, et al., "Hot-Electron-Induced MOSFET Degradation-Model, Monitor, and Improvement," *IEEE Transactions on Electron Devices,* Vol. ED32, No. 2, pp 375-382, February, 1985.

[3.3]   H. Peters, et al., "Compact SOI-MOSFET model for high temperature circuit simulation with emphasis on process and layout data," *Proceedings of the 1998 IEEE International Electron Devices Meeting*, pp 101-104.

[3.4]   K. Bernstein, et al., "High Speed CMOS Circuit Design Styles," *Kluwer Academic Publishers*, 1997, Chapter 1.

[3.5]   G. Shahidi, et al., "Partially-depleted SOI technology for digital logic," *Proceedings of 1999 IEEE International Solid State Circuits Conference*, February, 1999, pp. 426-427.

[3.6]   J. Eckhardt, et al., "A SOI-Specific PLL for 1 GHz Microprocessors in 0.25 μm 1.8V CMOS," *Proceedings of 1999 IEEE International Solid State Circuits Conference*, February, 1999, pp. 436-437.

[3.7]   R. Troutman, et al., *"Latchup in CMOS Technology,"* Kluwer Academic Press, 1986 ISBN 0-89838-215-7.

[3.8]   C. Adams, T. Kueper, IBM Rochester, Minnesota, informal discussions.

CHAPTER 4   *Static Circuit Design Response*

4.1 Introduction

This chapter will present the unique circuit responses that are present with partially depleted SOI in static circuits. In particular, the inverter, NAND and passgate structure will be investigated. The history effect, noise margins and bipolar current will be discussed with respect to these circuits. The effects of leakage and reduced diffusion capacitance will be covered. To open this chapter, let's first discuss the common SOI parameters that are important to the design point of SOI circuits.

4.2 Parameters of Interest to Circuit Designers

It is useful to begin our discussion by briefly reviewing the parameters asserted by SOI technology which are of interest to circuit designers. These include the capacitance of circuits, the threshold voltage of the device and the body effect.

1. Lower fan-out capacitance
   With source and drain diffusions extending down to the bottom of the SOI mesa,

the area component of the junction capacitance is greatly reduced. Since the BOX is much thicker than the depletion region of a bulk drain to substrate junction, the area component of the junction capacitance is almost negligible. The majority of the capacitance on the source/drain arises from lateral junctions facing the body. For partially depleted SOI, the lateral capacitance can be higher than the corresponding bulk capacitance since the source and drain are often more heavily doped to increase the leakage of the junction, thus resulting in a smaller depletion region and increased lateral capacitance. Depending upon the voltage of the body, it is possible for the capacitance of an SOI junction to be higher than if the junction was on a bulk wafer. This will be discussed further in Chapter 6 with the discussion of SRAMs. The capacitance of the gate to the body and inversion channel is nearly identical to a bulk gate capacitance.

The source/drain junction may couple to the substrate, but the capacitor's dielectric is the BOX and depending upon the thickness of the BOX, the coupling is likely to be very small.

Rule of thumb: Coupling to the substrate can be ignored for digital circuits on SOI.

2. Reduced Short Channel Effect
It has been previously mentioned in Section 3.4.7, that an elevated substrate bias results in less roll-off of device threshold voltage at very short channel lengths. This lack of reduction can create less performance gain as the channel length of the device is shortened compared to bulk. On a positive note, the leakage does not increase as quickly for SOI as the FET's length is shortened (see See 3.4.7, Reduced Short Channel Effect on page 46).

3. Lower Device Threshold
Without as much *Short Channel Effect* to worry about, V_T could be set lower to enjoy improved performance at all channel lengths. There is room for large potential performance gains at nominal channel lengths with reduced threshold voltages. One could take a different design point and set the threshold voltage higher for reduced subthreshold leakage.

4. Decreased Body Effect
Referring to Figure 4.1 a bulk NFET pulldown circuit is shown. Any voltage drop across device T1 elevates the potential of the source of transistor T2's above GND while current is being passed. T2, in absolute terms, sees a negative substrate voltage with respect to its source (voltage V_{ssx2} is negative), making T2's channel region harder to invert. Higher source-to-substrate bias V_{ssx2} increases the threshold voltage V_{T2}, and the circuit loses performance.

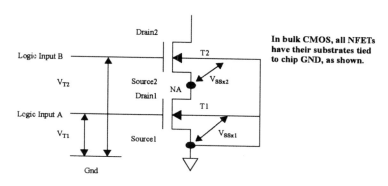

FIGURE 4.1 Body Effect

SOI does not have the body of the device attached to a common substrate. The body of transistor T2 will drift above ground and create an "Inverse Source Follower Effect" (body effect) which reduces voltage V_{ssx2}. With partially depleted SOI, the body will float to a potential that is between the source and drain potentials resulting in a negative V_{ssx2}, and leading to a reduction of the threshold voltage and improvement of the stacked device performance. This is not usually possible in bulk, therefore the name "Inverse Source Follower Effect" is given to SOI's body effect.

Each of these items are important at the device level and impact the performance. With a floating body, the need to consider the body in more detail. The next sections will explicitly describe the reaction of the body for an NFET in an inverter.

4.3 First Switch vs. Second Switch

Where did that 4th terminal come from?

In bulk CMOS technologies, we rarely consider the body as a fourth terminal for most circuits. We typically only consider the gate, drain and source of an FET to determine the electrical response of the circuit. Usually, the body of the FET is tied to ground for the NFETs and V_{DD} for the PFETs. As noted in Chapter 2, the body of an SOI FET is floating unless one intentionally adds a body contact to control the voltage. Without the body contact, the body is free-floating and is easily moved by the capacitive coupling to the gate, source or drain. Another competing factor for the control of the body is the two diodes that are formed from the body to the source and drain. If the FET is not switching, then the capacitive coupling is not a factor and the forward bias or reverse bias leakage of the diodes controls the body voltage.

An inverter's delay depends in part upon the threshold voltage of its devices. The threshold voltage is a function of the terminal voltages on the FET. Figure 3.5 on page 35 in Chapter 3 shows the effect on the threshold voltage as a function of the V_{BS}. As the voltage on the body of an NFET is increased, the threshold voltage decreases. This decrease in V_T can result in significant delay variations for different body voltages. With the variation of the body voltage and threshold voltage, the delay is dependent upon the potential of the terminals at the time that the input signal is received. This means the recent history of the terminals have an impact upon the performance of the circuit. This is known as the history effect and was predicted in [4.2]. The state of the terminals and the recent history also has an effect upon the noise margin and bipolar currents in pull down stacks, passgates and SRAM arrays.

Before we start, let us define some terms for usage during the rest of this book.

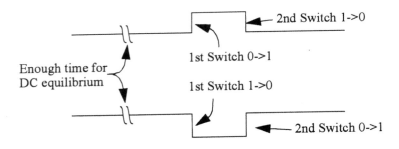

FIGURE 4.2 First switch vs. Second switch definitions.

Figure 4.2 shows two input signals, one starting at GND and one at V_{DD}. If each signal is held constant for a long time, then the circuit has reached DC equilibrium. The first transition will be called the 1st switch and depending upon the direction that the input signal switches, will be followed by the direction (e.g. 1st switch 0->1). One often wants to compare the first switch to the second switch for the same direction. To do this, two different simulations need to be done to capture the 1st switch 0->1 and 2nd switch 0->1. Now lets look at an inverter.

Figure 4.3 shows the switching characteristic for inverter. When the output of the inverter starts low, the V_{DS} of the NFET is 0V. The voltage of the body will come to a DC solution of 0V. As the inverter starts to switch, the input to the inverter falls. This is called the *first switch with the output starting low* or 1st switch 1->0. Since the body is lightly coupled to the gate, its voltage will dip below 0V slightly, and then the body's voltage will rise to 0.45V. This is due to the body being pulled high by the capacitive coupling between the body and drain. The pull comes from the PFET's drive on the output. This is an AC level. If no more switching were to take place, the body voltage would move to the equilibrium point of the diodes of the NFETs. This will be discussed later in the next section. Now, with the body voltage at 0.45V, the threshold voltage of the NFET is low. As the input transitions high, the body voltage will initially couple up from the capacitance between the gate and the body[1]. As the drain is pulled low by the NFET, the body will be capacitively coupled low, returning nearly to its previous level of 0V. This is called the *second switch when the output is high* or 2nd switch 0->1.

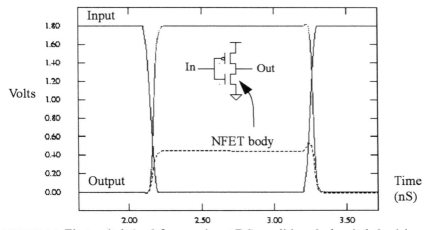

FIGURE 4.3 First switch 1->0 from an input DC condition. 2nd switch 0->1 is immediately following the first switch

Now, let's consider the case when the output of the inverter is at V_{DD} and the body voltage has come to equilibrium. The equilibrium point is a balance between the reverse leakage current from the diode across the body to drain junction and the for-

---

1. This upward coupling on the body caused by a rising gate depresses the threshold of the device and increasing current. This phenomena is referred to as *drain current overshoot* [4.3].

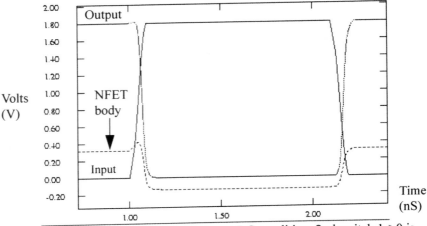

Volts
(V)

FIGURE 4.4 First switch 0->1 from an input DC condition. 2nd switch 1->0 is immediately following the first switch

ward bias current of the diode across the body to source junction. If this leakage is large, the body voltage will be further up the I-V curve of a diode and be lower, if the current leakage is small, the DC solution will establish a larger voltage on the body. Remember that the reverse leakage current for a diode is highly temperature dependent. Therefore, the DC body voltage will also be temperature dependent. In Figure 4.4 the body voltage is near 0.35V. As the input of the inverter switches high, the body initially couples upward due to the capacitance to the gate (see footnote on page 71) and then is pulled low due to the capacitance coupled to drain. The final voltage is near -0.15V. This transition is called the *first switch with the output starting high* or 1st switch 0->1. With the body below GND, the threshold voltage of the NFET is now larger than in bulk, where the body is held at GND. This is not a speed issue since then next transition is dominated by the PFET. Again, this is an AC condition only. If the input now transitions low, the body will not be pulled low since the channel shields the body from the gate. The body will start to move upward due to the coupling from the drain as the drain is pulled high by the PFET. The body voltage will again return nearly to its initial state since the capacitance is nearly identical between the first and second switch. This transition is called the *second switch with the output low* or 2nd switch 1->0.

Let's consider the timing of the two switch transitions:

First switch 0->1 and 2nd switch 0->1. During the first switch, the body started at about 0.35V. This resulted in a faster than bulk delay since the body of the NFET was above GND.

Second switch from the high output. Now consider the second switch from the high output. The body in this condition was near 0.45V and the threshold is even lower than the first switch. Therefore, the delay is even faster than the first switch.

Both delays are faster than an equivalent bulk circuit, but the delays were not equal. The first switch was the slowest SOI switch, the second switch was the fastest. Therefore, the delay of two identical circuits depends on previous state of the output of the circuit and has a history effect. One gets the same answer for the opposite set of switches from 1->0 that the 2nd switch is faster. This is not a rule of thumb, though. The 2nd switch can be made slower than the first switch by making the DC body voltage higher than the AC body voltage. One method of achieving this is by reducing the leakage of the diodes to the body. Another method for changing the relation of 1st switch to 2nd switch is by changing the amount of capacitance between the body and the drain. If the body movement due to coupling is reduced, then the 1st and 2nd switches will vary in delay[4.6]. To have no history effect, one wants to have the amount of voltage swing due to capacitive coupling to equal the DC equilibrium voltage when V_{DS} is equal to V_{DD}. This may be attainable at one temperature. However, since diode leakage is strongly temperature dependent, it cannot be maintained over a product's entire operating range. Another way to produce minimal history effect is to change the Wp/Wn ratio of the devices. Instead of using a more traditional 2:1 ratio, a ratio of 1:1 will produce little to no history effect[4.7]. This method has again balanced the amount of capacitive body coupling by changing the slew rate on the drain of the NFET to the DC equilibrium potential. This is only true at one temperature as the DC equilibrium is strongly temperature dependent.

One artifact of having asymmetric delays such as the first switch being slower than the second switch is that a rising pulse will shrink in width as it propagates down a delay chain. Therefore, there is a minimum pulse width for a given inverter chain length. The longer the chain, the wider the pulse width needs to be for the pulse to exit the chain without completely collapsing.

So far we have been dealing with the response of an NFET's floating body. As one can imagine, there is a symmetry to the PFET's response. A PFET's body potential is usually tied to V_{DD} in an NWELL. However, when it is left floating, acts in a similar manner. As the body floats below V_{DD}, the magnitude of the threshold voltage of the PFET is decreased and results in improved performance. As is the case with the NFET, it is possible for the body of the PFET to go above the supply rail with coupling from the drain when the inverter's output initially started low. We will continue the discussions in the rest of this chapter by working with the NFET and remembering that a PFET has a similar response.

Figure 4.5 shows the same voltage characteristics as Figure 4.4 with the exception

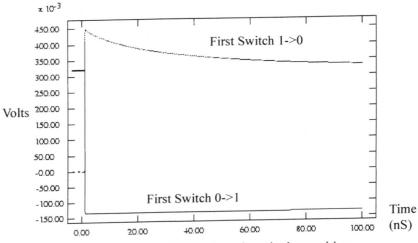

FIGURE 4.5 Body voltages of an NFET with only a single transition

that there is no second switch. Notice that the body voltage that started at GND and was coupled to 0.45V did not stay there. After the first switch, the drain of the NFET is at V_{DD} and the source is at GND. The body to source junction is a forward biased diode and will dissipate the charge in about 100 nanoseconds.

In a similar fashion, consider the body voltage that started at the equilibrium point of 0.35V. After its first transition, the body voltage was near -0.15V. Here, both of the diodes that are connected to the body are reversed biased and only the reverse bias leakage current can add to the charge and increase the voltage. This is highly process dependent upon the doping profiles on both sides of the junction. In any case, the DC equilibrium point of body will eventually return to 0V, but may be as long as several hundred milliseconds.

4.4 First Switch vs. Steady State

It has been shown that the speed of a circuit depends upon the body voltage of the FETs and the body voltage depends upon it most recent switching history. Most circuits will switch more than once within several nanoseconds and we cannot assume that the circuit will always return to the DC condition. Figure 4.6 shows the body voltages of an NFET starting from the two DC states of output high and output low..

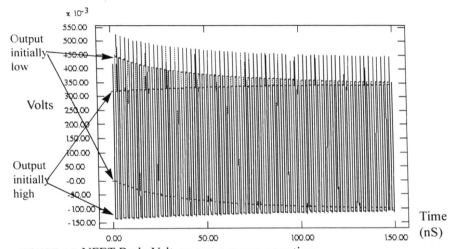

FIGURE 4.6 NFET Body Voltage convergence over time

This graph is rather hard to read, but notice that over time the body voltages converge to a steady state voltage that lies between the DC value and the 2nd switch value. Thus, the delay for the nth switch and nth+1 switch do not vary much when n is a large number. The steady state point for this circuit which had a 50% duty cycle results in the body voltages at nearly 0.35V when the drain is high, and -0.1V when the drain is low. This steady state position is *faster* than the first switch, but *slower* than the second switch. The delay for a free running circuit is easily predictable provided that it is *constantly* switching with the same duty cycle. Most circuits do not switch every cycle, nor do they sit at equilibrium for long periods of time. Since the steady state delay lands somewhere between the first and second switch delay, random switching will result in a delay somewhere between them as well.

One interesting point is that the convergence to steady state was faster for the 1st switch 0->1 than under a single transition shown in Figure 4.5. Since the transitions sometimes switch the body to source diode from reversed biased to forward biased, the decay rate is faster than if it had stayed in the reversed biased state all of the time.

Since we have talked about the history effect in terms of 1st switch, 2nd switch and steady state, how does one put a number to the amount of history that is present in a circuit? For this text, we will declare *history* to be defined as the percentage change in delay of the circuit between the 1st and 2nd switch.

4.5 Static Circuit Response to SOI

I am an EE, I was told there would be no history...

Static circuits are very robust circuits for all types of applications. Since static circuits maintain an active drive on the output, they are robust and relatively immune to noise [4.1] . In this section, we will discuss the inverter and the 2 input NAND. We have already discussed the inverter with respect to the switching capabilities and body voltages of NFET in the previous section. The biggest advantage of SOI across the different static circuits is the speed improvement, especially as stack heights are raised.

4.5.1 Inverter Response

We have already spent some time discussing inverters with the presentation of the first, second and steady state switch points. Figure 4.7 shows the delay of an inverter with a fan out of 3 in bulk, in SOI with floating bodies and in SOI with a body contact connected to every device and the body tied to V_{DD} or GND for the PFET and NFET, respectively. At a fan out of 3 with no wire delay, the average SOI delay is 20-30% faster than the bulk delay. The speed improvement for inverters comes from the reduction of the depletion capacitance on the junction and the inverse body effect. In a bulk inverter, the V_{BS} is normally zero, but in SOI, the V_{BS} is positive resulting in a lower threshold voltage and thus a smaller delay. The final curve in Figure 4.7 is with a body contact on every stage of the inverter. Here, the inverse body effect is resolved and the first and second switch variation is reduced. The gain for the body-contacted SOI device is now solely due to the removal of the junction capacitance. However, the increase in the gate capacitance caused by the large body contact and the extension to the gate resulted in more input capacitance than the floating body inverter. Overall, the body contact was detrimental to the inverter's performance. If body contacts become smaller or cause less performance degradation, then they may be used more regularly for static logic. There are several circuits that require body contacts. These

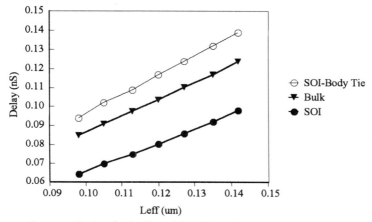

FIGURE 4.7 Inverter Delay for bulk and SOI. Two SOI curves are shown, one with floating bodies and one with contacts to the body.

include analog circuits, I/Os, and locations where the delay variability cannot be tolerated.

> **Rule of thumb:** Use body contacts sparingly, on analog circuits and where history effects are not tolerable.

It has been shown that the body voltage is dependent upon the previous history of circuits. This directly correlates to the delay of the inverter being dependent upon a combination of recent history and "long term" history of its inputs and outputs. Figure 4.8 shows the delay of an inverter for the 4 options of 1st and 2nd switch starting at 0 or 1. Here the input pulse is a constant width and a constant frequency. Again, the second switch is shown to be faster than the first switch. The legends are a bit misleading as the top (slower) curve is really the 1st, 3rd, 5th... switch and the bottom (faster) curve is the 2nd, 4th, 6th... switch.

The steady state delay is achieved after several hundred nanoseconds. It lies between the fastest (2nd switch) and the slowest (1st switch) delay. One can see that the 0->1 decay to the steady state value occurs more quickly than the 1->0 decay to its steady state value. This is due to the fact that the decay rate is determined by the discharge of the body potential; the faster convergence to steady state is a result of the forward biased diode and the slow curve is due to a reversed biased diode.

The history effect as a function of pulse width to the second switch is shown in Figure 4.9. As one can see, the frequency and the voltage determine the amount of history effect. The slower the frequency, the less the history effect, since the body

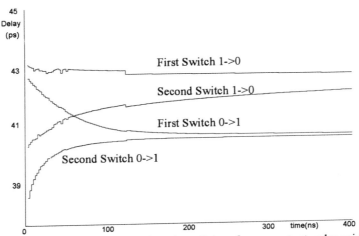

FIGURE 4.8 Inverter delay as a function of time for a constant pulse width and frequency.

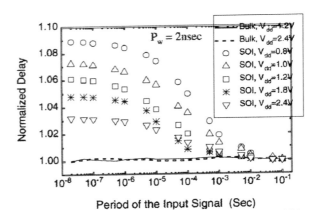

Period of the Input Signal (Sec)

FIGURE 4.9 Inverter delay as a function of the input pulse width.

potential has had time to adjust back to a value closer to its DC equilibrium point. The equilibrium point occurs between 10mS and 100mS. Looking at Figure 4.5 on page 74, one can see that coming out of DC equilibrium, the difference in the body voltage is about 320mV. Immediately after the transitions, the difference rises to near 600mV and starts to decay. The longer one waits, as in a slower frequency, the closer the second switch delay is going to be to the first switch's delay.

4.5.2 Noise Response for Inverters

With lower threshold voltages on the devices, one would suspect that the noise issues for an inverter in SOI may be worse. At the same time, the current carrying capabilities are better for each device, and will result in an inverter being able to suppress a bit of noise on the input. These two competing factors will determine the noise margins. Figure 4.10 shows the transfer curve for an inverter in bulk and SOI. Bulk has a

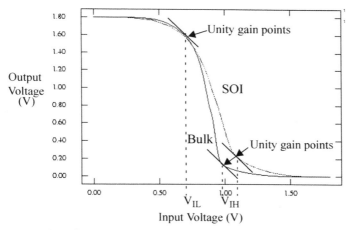

FIGURE 4.10 Transfer curves for an inverter in bulk and SOI.

steeper curve. Since the device size did not change and the curve has shifted to the right this demonstrates that the PFET has gotten stronger than the NFET compared to bulk inverter, for this implementation of SOI. As the input voltage starts to rise, the first unity gain point is nearly identical for both curves, resulting in nearly identical input low noise margins (N_{IL}). However, the slope of the bulk curve is steeper and arrives at a the second unity gain point at a lower input voltage than SOI resulting in the bulk inverter switching sooner than SOI inverter. Thus, the input high noise margin, $N_{IH}=V_{OH}-V_{IH}$, is lower for SOI than for bulk. Being a CMOS inverter, the output range is not compromised and still switches from GND to V_{DD}. The definitions of noise margin can be found in [4.4].

How does the difference in the unity gain points translate to the possibility of an incoming noise pulse being transmitted through an inverter? Figure 4.11 shows a noise schmoo for the same inverter as in Figure 4.10. For this plot, an upward pulse was presented at the input of the inverter with varying pulse width and pulse height.

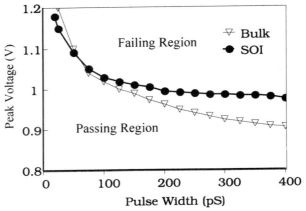

FIGURE 4.11 Noise Schmoo for an inverter in SOI and bulk CMOS.

The upper region is the failing region, where a fail is defined as the output of the inverter passing $V_{DD}/2$. As the pulse gets wider, it requires less pulse height to cause the inverter to pass the input pulse onto the output and produce a "glitch" or a fail. Even though the noise margin, N_{IH}, is lower for SOI than for bulk, the SOI's noise schmoo for this inverter is a bit higher and more resistant than bulk to an input pulse that is wide, but slightly worse as narrow input pulses.

A similar plot could be made with a downward pulse. Since the input low noise margin (N_{IL}) is nearly identical, the noise schmoo would be nearly identical.

4.5.3 NAND Gate Response

Let's do something more than an inverter.

A NAND gate demonstrates more of the SOI advantages for speed. In a stacked configuration like the NFETs in a NAND structure, the FETs in a bulk technology not attached to GND are subject to more body effects and increase the delay. In SOI with the floating body, these FETs now are free to operate in the inverse body effect range. Unlike an inverter, where body voltage of the NFET was limited to 0.32V under DC conditions, the body voltage of the top devices in the stack can be as high as V_{DD}. This occurs when the output of the NAND is high in response to all devices in the NFET pulldown tree are off. The top device in the stack may have a potential near V_{DD} on the source and V_{DD} on the drain of the device and the body voltage will come to an equilibrium point near V_{DD}.

Figure 4.12shows the response of a two input NAND gate for the first and second

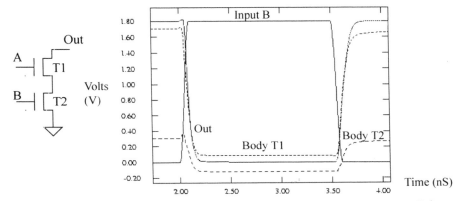

FIGURE 4.12 Schematic and simulation for a two input NAND gate. Input B is switched with Input A tied to V_{DD}. Body voltage for T1 and T2 are shown.

switch with the input signal arriving at input B and input A is tied to V_{DD}. The body voltage of transistor T2 starts off at 0.32V just like an inverter. Upon the transition of input B, the body voltage of T2 dips below ground and is pulled back to 0.32V when the input returns to GND its drain returns to near V_{DD}. Again, this action of the body of T2 is like an inverter.

The body of the transistor T1 starts very near V_{DD}. It does not start all the way up at V_{DD} since the node between transistors T1 and T2 is not precharged and comes to a DC value of V_{DD}-V_{Tbody} (not shown). If the node between the transistors had been precharged to V_{DD}, the body voltage of transistor T1 would have been V_{DD}. In this figure, the body voltage started at 1.7V. Now when the input goes high, the body of the top device is pulled low by the capacitive coupling of the source to body and the drain to body capacitance. Charge is drained through the forward biased diode that is the body to source junction and further reduces the body voltage. The body voltage settles near 0.1V. The body voltage of T1 will drift slowly towards 0V since both the source and drain of this FET is at GND. When input B returns low, the body voltage of T1 is pulled high with coupling capacitance from both the source and drain. The body will not immediately return to its DC equilibrium value since there is not enough coupling to do so. Notice that if another switch was to occur at 4.0nS, the delay of the NAND would be slightly slower than the first switch since the body voltage is lower than the DC equilibrium value.

We are having too much fun! Now, let's consider the same NAND2 circuit, but with input A tied to GND. Figure 4.13 shows the same transitions on input B, and the body

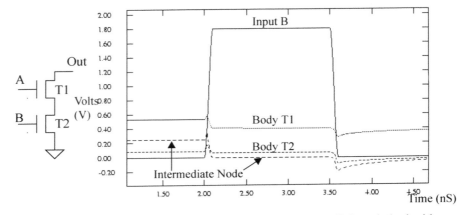

FIGURE 4.13 Simulation for a two input NAND gate. Input B is switched with Input A tied to GND. Body voltage for T1 and T2 are shown.

of the two FETs. The voltage on the intermediate node between T1 and T2 is shown, as it has a large impact upon the body voltage of T1. Since the output always stays at V_{DD}, it is not shown. At DC equilibrium, the intermediate node voltage is at a value balancing the subthreshold leakage currents between transistors T1 and T2. This allows the body of T1 and T2 to start at an equilibrium voltage of 0.55V and 0.10V, respectively. After input B has transitioned high, the intermediate node initially is pulled up by the Miller capacitance of T2 and then is pulled to ground as T2 turns on. The potential on the body of T2 is also initially coupled upward due to gate to body capacitance and then is pulled low by the intermediate node. Eventually, it will drift toward ground since both the source and drain junctions are slightly forward biased. The body of T1 is coupled upward slightly since its source, the intermediate node moves up and then is pulled down when the intermediate node goes to ground. The body of T1 goes to a potential of 0.4V. The second transition on input B doesn't have much effect on the intermediate node nor the body potentials. One can see in this figure that the input coupling from the gate of the device does cause a blip in the potential of the bodies. After the second transition, the bodies will begin to drift back to the equilibrium voltage.

Now for some interesting plots. Lets assume that inputs A and B arrive at the same time and switch as input B did before. Figure 4.14 shows the complexity that is possible if one tries to keep up with the body voltage under all conditions. For static cir-

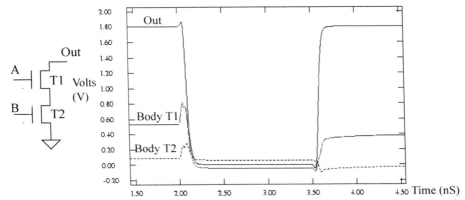

FIGURE 4.14 Schematic and simulation for a two input NAND gate. Inputs A and B are switched at the same time. Body voltage for T1 and T2 are shown.

cuits, it is not necessary to watch the body voltage of each device since their potentials are limited to about a V_{be} above the source[2]. The effect is to dynamically modulate the threshold voltage of the devices and slightly increase or decrease the device currents. The DC equilibrium voltage is identical to the previous scenario. Once the input switches, the body voltage of T1 is pulled in several directions before it ultimately falls to -0.05V. The voltage on the body of T2 behave nearly the same as it did in the previous simulation.

Now that it has been shown that the body potential is much different than bulk, what has it done to my speed? At a fan out of 3 with no wire delay, the delay on this two input NAND structure is 30% faster than bulk. Figure 4.15 shows the delay improvement for SOI over bulk for varying stack heights, without body contacts. The pulldown height of one is the inverter discussed in the previous section. The more NFETs in the pulldown tree of a NAND gate, the bigger the delay improvement over bulk. This leads one to the conclusion that it would be wiser for an SOI design to use a pulldown network with a larger number of stacked device so one may reduce the number of logical gates in a circuit to improve the speed.

Since a NAND gate has several stacked devices in its pulldown tree, the last arriving signal should arrive at the gate that will provide the fastest output transition. In SOI, the input ordering has not changed. As in bulk, the last arriving signal should still

2. V_{be} is the parasitic bipolar devices base-to-emitter voltage.

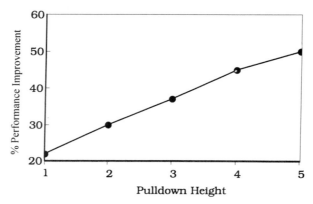

FIGURE 4.15 Delay improvement vs. NAND stack height.

arrive at the top device in the stack. Even though the upper devices in the pulldown tree have improved more in SOI than in bulk, the input ordering has not changed.

Though not discussed here, the same argument for NAND gates can be applied to NOR gates. The PFET's inverse body effect improves the delay of NOR gates. The larger the stack height in the pullup tree, the greater the improvement of the delay for an SOI circuit over its bulk counterparts. For a table of SOI's performance gains over bulk see Figure 5.19, "SOI performance improvement over bulk for a recent design [5.2]," on page 116

4.6 Passgate Circuit Response

Beware of the bipolar current.

Passgate circuits are a convenient way to control the path of a signal and not induce as much of a delay penalty as a stacked gate such as a "tri-buff". The use of passgates in muxes and latches are very beneficial and straightforward. Figure 4.16 shows a few options for schematics of passgates. The passgate circuit itself either comprises a NFET only passgate (A) or a complementary CMOS transmission gate with both an NFET and a PFET (B). The loads of the circuit are typically either a full latch (C), a half latch (D), or an inverter (E). For this discussion, the intermediate node is the net between the passgate and the load. In this section, we will analyze an NFET passgate

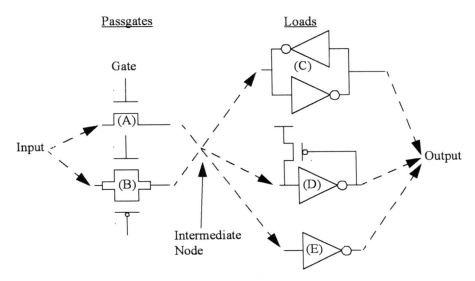

FIGURE 4.16 Passgate schematics.

with no load, an inverter load (A-E) and a half latch load (A-D). The ideas presented here are extendable to the other possible configurations.

With SOI's floating body, the body of the passgate is the one to watch. Let's take a look at several of the scenarios when the passgate is off. In these cases, the gate is at GND, input will be at V_{DD} and intermediate node will start at V_{DD}.

When the input transitions from V_{DD} to GND, an SOI-unique situation occurs. This scenario is avoided in bulk by tying the substrate to GND and the NWELL to V_{DD}. In SOI, the body of the NFET is above its source by more than a diode drop. Therefore, the body to source diode is forward biased and the body to drain diode is reversed biased. This action turns on the parasitic NPN bipolar device that is in parallel with the NFET. Since the body (or in this case the base) of the bipolar transistor is not actively driven, the current through the NPN bipolar transistor only lasts as long as is required to remove the charge from the body. Figure 4.17 shows input transition, the body voltage and the at the same time the bipolar current is shown. In this figure, the bipolar current is magnified 100 times and the movement of the body will be discussed more in a bit.

The height and width of the bipolar current is dependent upon the design of the device. If the gain of the parasitic NPN transistor is large, the pulse height will be large and pull more charge from the intermediate node. If the body of the FET has lots

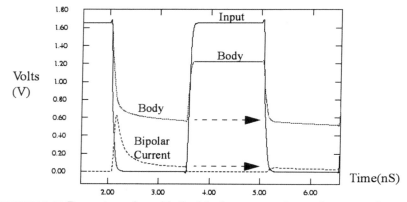

FIGURE 4.17 Passgate nodes with the bipolar current shown (current values not shown).

of charge to be removed, the width of the current pulse will be larger and again, a large amount of charge may be pulled from the intermediate node. Also, at increased supply voltages, the capacitor divider couples the body voltage down less, leaving more charge to be drained by the parasitic bipolar transistor. In any case, since the body of the FET is floating and no current is being actively driven into the base of the parasitic NPN transistor, it is impossible to maintain the bipolar current.

One interesting aspect of the body voltage and bipolar current is also shown in Figure 4.17. If the input is transitioned faster than the decay of the body voltage, the body voltage will return to nearly the same voltage as where it had left off (see arrow). This also results in a continuation of the bipolar current charge loss across several cycles, in a "pumping action". This continuation needs to be accounted for if the holding circuit releases the intermediate node on a subsequent cycle.

With different loads on the output, the intermediate node will have varying responses. The three cases that will be considered are a passgate with no load on the intermediate node, a passgate with only an inverter attached to the intermediate node and a passgate with a half latch attached to the intermediate node.

In all three cases, given enough time, the body voltage of a passgate circuit will float up to V_{DD} as shown on the left hand side of Figure 4.18.

Pass Gate, No Load

As the input is pulled low, the body is also pulled low. The source to body capacitance couples the body downward to a voltage determined by the capacitive divider. This

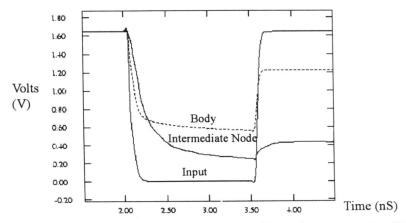

FIGURE 4.18 Passgate response to drain being pulled low when passgate is off. No load is present on the intermediate node.

will usually be slightly above $V_{DD}/2$. If $V_{DD}/2$ is above V_{be} then the body to source diode becomes forward biased and drains charge from the body. The parasitic bipolar current pulls the output low. Since no holding circuit and very little capacitance sits on the intermediate node, the intermediate node potential is pulled very low. If the device is left at this point, the body of the passgate would settle out at the same voltage as an "off" NFET in an inverter at 0.32V. However, in this simulation, the input is pulled back up to V_{DD}. The body does not return to V_{DD} since the capacitive coupling is not large enough. In addition, the charge in the body has been drained due to the forward biased diode. The body only returns to 1.2V. The intermediate node couples only slightly back upwards due to the series capacitance from the source to the drain. Given enough time, the body and the drain would return to V_{DD}, but this is very slow, since the body and the intermediate node would be pulled up by reverse bias leakage across the two diodes that are connected to the body. It may take up to one second to return to V_{DD}. For a free-running source with the passgate off, the body voltage will reach a steady state switching between 0.32V and 0.9V. This is not shown in this graph.

Pass Gate Driving an Inverter

Here, the passgate now has added an inverter on the intermediate node (A-E in Figure 4.16). The addition of the inverter adds capacitance to the intermediate node, but does not actively help to hold or move the potential on the intermediate node. Figure 4.19 shows the same input switching as the previous scenario. In this figure,

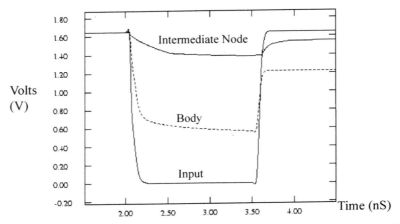

FIGURE 4.19 Passgate response to drain being pulled low when passgate is off. An inverter is attached to the intermediate node.

the body starts at V_{DD} again and as the input switches low, the body again is pulled low due to the source to body capacitance. Again, a bipolar current passes through the parasitic bipolar device and pulls the intermediate node low. Since there is more capacitance upon the intermediate node due to the addition of the inverter as a load, the amount that the intermediate node is pulled low is much less than in Figure 4.18. As the input rises, the body and the intermediate nodes also capacitively couple upwards, but neither approach the rail in this short time period shown. Both the body and the intermediate node would reach V_{DD} given enough time.

The authors do not recommend the use of an NFET passgate with an inverter load only (A-E) as the intermediate node is not held by an on FET when the passgate is off and does not maintain a V_{DD} level during typical cycle times in today's microprocessors.

Pass Gate Driving a Half-latch

The load on the intermediate node is now a half latch (A-D in Figure 4.16). The goal with the half latch is to help hold the intermediate node to voltage V_{DD}. With this half-latch configuration, the intermediate node should not be disturbed except by the bipolar current. Figure 4.20. shows the same input transitions as the previous two scenarios. In this case, one can see that the intermediate node did dip a bit as the source transitions from V_{DD} to GND, but the feedback PFET was sized such that it could absorb the parasitic bipolar current and hold the voltage on the intermediate node.

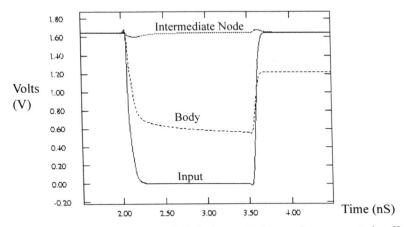

FIGURE 4.20 Passgate response to drain being pulled low when passgate is off. A half latch is attached to the intermediate node.

The body voltage is nearly identical to the other scenarios, since this voltage is dominated by the forward biased diode on the source.

If the bipolar current is large enough, the data on the intermediate node may be corrupted even with a half latch, if the passgate and the feedback PFET are not correctly sized. This failure mode is possible in circuits such as a latch or a tristate bus, and is known as the "unbuffered passgate" controlled storage hazard. This is a new side of a common failure mode which should be familiar to designers of submicron bulk circuits. In bulk, the parasitic bipolar device can be activated if the input node is coupled below beyond the power supply by a V_{be}. Here, the failure can occur in the absence of coupling. Consider Table 4-1 which lists four possible noise scenarios for a complementary passgate followed by a latch or a half-latch. The voltages listed are for the DC equilibrium case *before* a noise event arrives. For bipolar current, the input potential must transition completely from rail to rail. For coupling noise, the coupling from the neighbor causes the input to go below GND or above V_{DD}. The final two columns list the possible fail mechanisms for the complementary passgates and under which conditions noise is a problem. In these two columns, the device that is providing the undesirable current flow is also listed. For case #1, the PFET would provide the parasitic bipolar current. Case #4 is the situation that we described earlier in this section. For the coupled noise cases in case #2, the PFET provides the current when it turns on and the NFET provides the current for case #3. In bulk, adding an inverter before the

Case	V_{Input}	$V_{Intermediate}$	V_{Body}	Bipolar?	Coupling Noise?
#1	GND	GND	GND	Yes PFET/PNP	No
#2	GND	V_{DD}	0.35	No	Yes NFET
#3	V_{DD}	GND	0.35	No	Yes PFET
#4	V_{DD}	V_{DD}	V_{DD}	Yes NFET/NPN	No

TABLE 4-1 Table of Potential Noise issues for passgates in SOI

passgate or changing the passgate to a tri-buf will eliminate the problem. One could also decouple the wires by changing physical proximity to aggressor wires. Passgate noise is discussed more in [4.1]. In SOI, there is not a direct way to prevent the current unless one can prevent the transition at the input of "off" passgates. Otherwise, one must design to make the circuit tolerant of the bipolar currents.

Methods for reducing the bipolar current are often at the expense of speed.

1. To reduce the size of the bipolar current one may simply reduce the width of the passgate. This may not be desirable if the feedback is already strong and may cause the half latch or the latch to be unwrite-able.

2. One may strengthen the feedback of the device in the half latch or full latch configuration. This will not reduce the bipolar current, but it will decrease the magnitude of the voltage drop on the intermediate node, which will reduce the possibility of noise being propagated to the output of the latch.

3. One may reduce the switch point for the forward inverter in the load portion of the circuit. Again, this will not reduce the bipolar current, nor will it decrease the voltage drop on the intermediate node, but it will diminish the amount of the noise propagated to the feedback gates through the output inverter.

For all individual SOI passgates, the authors recommend that some form of feedback be provided to help hold the intermediate node.

This passgate "leakage" current is particularly a problem for wide muxes that have many NFETs in parallel. For example, a 32-to-1 mux may have 32 NFETs connected to the intermediate node. If all of them were off and all 32 inputs transitioned from V_{DD} to GND, then the intermediate node would be discharged by 32 bipolar currents. Simulations have shown that in some design corners, the number of passgates on a mux may be has high as 16[4.8]. It is possible not to have a solution to the problem. If one makes the passgates small enough that all 32 combined will not corrupt the intermediate node, it may not be able to override the strength of the feedback devices and would fail to write the latch.

Often in these wide mux configurations, one of the passgates is conducting as the selection of the mux is orthogonal or "one-hot". Feedback is not needed in this topology, since the conducting passgate provides the current to maintain the voltage level. In this scenario, bipolar currents can still create a large amount of noise and cause a temporary glitch on the output of the mux.[4.5]

Complementary passgates have been discussed in the previous paragraphs and the PFETs also will produce a parasitic bipolar current when the input transition from GND to V_{DD}, but in practice, the gain of the PNP transistor is much less than that of an NPN and is not thought to be a problem.

> **Rule of thumb:** Bipolar currents must be considered through NFET passgates, but not PFET passgates.

Bipolar currents will get more attention in the following chapters on dynamic circuits and SRAMs. Note that we did not mention the bipolar current when discussing the static circuits earlier in this chapter. It is indeed possible to produce bipolar currents. In stacked circuits such as the one shown in Figure 4.13, but since the output is always driven by the complementary device (in this case the PFETs), the bipolar current will not corrupt the NAND's output, and is inconsequential.

Let's investigate how a passgate works when it is on. Figure 4.21 shows the voltages of the input, body of the FET and the intermediate node when the gate of the passgate is at V_{DD} and the input toggles starting at V_{DD}. As before, at DC equilibrium, the body is at V_{DD}. When the input transitions to GND, the body is coupled by the capacitance to the input and the intermediate node. The coupling brings this down to about 0.15 V. When the input transitions back to V_{DD} within a few nanoseconds, the capacitance from the input and the intermediate node pulls the body all the way back to V_{DD}. Since the input transitioned back within a brief period, the body movement in this case is essentially due solely to capacitive coupling. That is why the body returns to its original potential. Given enough time, if the input had not transitioned back to V_{DD}, but stayed at GND, the body would have eventually reached GND since both junction diodes connected to the body are forward biased. Once the body had reached

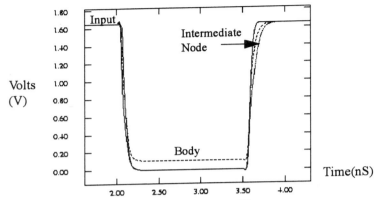

FIGURE 4.21 Passgate voltages with the gate at V_{DD}.

GND, the upward transition on the input would have pulled the body up to 1.65V due to the same capacitive coupling and would have had to "leak" back up to the DC equilibrium through the junction diodes connected to the body that are now reversed biased.

In either case, the body to source voltage is very small at all times for an on passgate. In bulk, the body would be at GND and the body to source voltage would initially be at $-V_{DD}$ resulting in a large body effected threshold voltage and limiting the performance. The performance improvement of passgates in SOI is substantial when compared to bulk: pass gates, in fact, provide a great opportunity for SOI performance improvements. Given safeguards to the issues described above, pass gates and SOI can go together like peanut butter and jelly.

4.7 Summary

Static circuits respond well to SOI. The performance of an SOI static combinatorial circuit is almost always faster. This motivates pursuit of implementing logic using static SOI circuits. Several SOI circuit responses are present, such as history effect and bipolar currents due to the floating body of partially depleted SOI. For the inverter, the body was straightforward to understand and follow quantitatively. Once a NAND structure was introduced, the possible combinations increased substantially and the exact position of the body was not as easy to predict except under DC equilibrium conditions. The stacked structures have a bipolar current possible, but the nature of static circuits did not result in a serious design issue.

For passgate circuits, the value and direction of the floating body as the gate, source or drain were switched are easy to understand. Here, the bipolar currents are prevalent and must be accounted for in all designs. Wide muxes or passgates without a very strong feedback path are prone to failing due to charge removed from the storage node by the bipolar current. Well-tuned passgate circuits can be used efficiently in SOI.

REFERENCES

[4.1] Kerry Bernstein, Keith Carrig, Christopher Durham, Patrick Hansen, David Hogenmiller, Edward Nowak, and Norman Rohrer, *High Speed CMOS Design Styles,* Boston, Kluwer Academic Publishers, 1998, pp. 160-166.

[4.2] Dongwook Suh and Jerry G. Fossum, *IEEE International Electron Device Meeting Technical Digest,* December 1994, pp.661.

[4.3] Y. Taur, et al., *Fundamentals of Modern VLSI Devices*, Cambridge, England, Cambridge University Press, 1998, pp. 280-282.

[4.4] Neil Weste and Kamran Eshraghian, *Principles of CMOS VLSI Design,* Reading, Massachusetts, Addison-Wesley, 1985, pp. 51-53.

[4.5] Pong-Fei Lu, Ching-Te Chuang, Jin Ji, Lawrence F. Wagner, Chang-Ming Hsieh, J. B. Kuang, Louis Lu-Chen Hsu, Mario, M. Pelella, Jr., Shao-Fu Sanford, and Carl J. Anderson, "Floating Body Effects in Partially Depleted SOI Circuits" *IEEE Journal of Solid-State Circuits*, Vol. 32, No.8, August 1997, pp. 1241-1253.

[4.6] Duckhyun Chang, Byoung Min, Surya Veeraraghavan, Michael Mendicino, Troy Cooper, Skip Egley, and Kevin Cox, "Temperature Dependent Hysteretic Propagation Delay in FB SOI Inverter Chain", *'roceedings of 1999 IEEE International SOI Conference,* October 1999, pp.82-83.

[4.7] M. M. Pelella, C. T. Chuang, C. Tretz, B.W.Curran, and M. G. Rosenfield, "Hysteresis in Floating-Body PD/SOI Circuits", *1999 International Symposium on VLSI Technology, Digest of Technical Papers,* June 1999, pp. 278-281.

[4.8] P. F. Lu, J. Ji, C. T. Chuang, L. F. Wagner, C. M. hsieh, J. B. Kuang, L. Hsu, M. M. Pelella, S. Chu, and C. J. Anderson, "Floating Body Effects in Partially-Depleted SOI CMOS Circuits", *International Symposium on Low Power Electronics and Design, Digest of Technical Papers 1996,* pp. 139-144.

CHAPTER 5

Dynamic Circuit Design Considerations

5.1 Introduction

SOI's favorable power and performance characteristics makes this technology appealing to a wide range of market applications. Its reduced capacitance and increased transconductance makes it attractive specifically to high performance applications. It follows, then, that it would be counterproductive to eliminate aggressive high speed circuit styles which may be at risk due to some of SOI's idiosyncrasies described in the last chapter. In this chapter, practices are discussed which help the dynamic logic circuit designer both reduce the temporal contribution to variation and avoid hazardous parasitic responses, without the need to eliminate high speed topologies from the design.

The origin of temporal variation is the varying amount of charge present in the floating body of an evaluation transistor. This charge influences the voltage in the body, which in turn affects the threshold voltage and hence device performance. The range of possible body voltages are different for each of the modes an SOI MOSFET in a dynamic circuit might operate in:

- In the DC case, the gate input to the transistor has been stable for a sufficiently long period to allow the device's body to converge on a stable balance of source/drain junction leakage and impact ionization current. This is commonly referred to as the "equilibrium state."

    Rule of thumb: With the gate and source low and drain high, the transistor body is at its highest DC potential, approximately 0.32 V for an NFET, and the transistor is pre-conditioned to operate with high performance.

Rule of thumb: With the gate high, and the drain and source low, the transistor body is at its lowest DC potential, 0 V for an NFET, and the transistor is preconditioned to operate with lower performance.

- During device transitions in high speed operations, the body voltage is strongly coupled to the drain, source and gate of the transistor. At various instances, the body will be most influenced by one of the three, and is violently coupled up and down in voltage by its movement. Large excursions in circuit performance may be observed in this mode, commonly referred to as the "dynamic state."

  Rule of thumb: Expect 8-10% delay variation from the Steady State case due to dynamic body coupling.

- In the "Steady State Case," a given SOI logic path has been running at a fixed frequency for a sufficient period such that the upper and lower limits to the body voltage are constant, and the delay of the path in each transition direction remains stable.

The goal, then, is for the designer to minimize the impact that these various states of operation have on dynamic circuit delay and reliability.

5.2 Dynamic Circuit Response

Dynamic circuits respond to SOI in much the same manner as static NAND gates and passgates. The difference lies in the inability of dynamic circuits to sustain charge loss. Bipolar currents were not a concern in static NAND gates, as they would not cause a fail. They are present, but with the output always being actively driven, the small bipolar current is easily replaced. With dynamic circuits, the bipolar current must be considered. As an example consider the following scenario:

5.2.1 Dynamic History Effect

In Figure 5.1, we examine a dynamic circuit with a use history that sets up the potential for bipolar current. Assume input A0 has been held high and inputs B0, C0, D0 have been held low for an extended period. With no path to GND, nodes N1 and N2 have remained charged to voltage V_{DD} and $(V_{DD}-V_T)$ respectively. Note that now the bodies of devices 3, 4, and 5 are charged to approximately V_{DD} even though only input A0 had been high. Assume now, that after another precharge interval, input C1 goes high, and inputs A0, B0, and C0 all remain low. As node N2 is discharged to

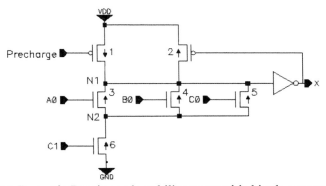

FIGURE 5.1 Dynamic Domino vulnerability to parasitic bipolar response.

GND, the bodies of devices 3, 4, and 5 discharge through their forward-biased body-to-source diodes. This body-to-source current "I" in each device initiates an additional bipolar drain-to-source current response of "βI" in each device which persists until the respective body is discharged. This current has the potential to discharge the intermediate node N1 sufficiently to invalidate signal output at node X. Figure 5.2 shows the simulated effect of bipolar currents on the stored dynamic precharge in the domino circuit of Figure 5.1. In this plot, device 3 had been left on during precharge, charging node N1 to V_{DD} and node N2 to V_{DD}-V_T. Precharge clock ("Clock") goes high to initiate evaluate, and device 6 is turned on shortly after and pulls node N2 to GND. Note that although the inputs to device 3, 4, and 5 are all low, the voltage of node N1 droops due to the bipolar currents passed by devices 3, 4, and 5. Node N1's droop causes a noticeable temporary rise in output X. Section 5.3 will discuss methods for designing around the history effect and bipolar currents.

5.2.2 Dynamic Charge Sharing

Charge Sharing refers to the capacitive division of precharge on the summand node of a domino circuit to other intermediate nodes of the circuit. Referring once again to Figure 5.1, consider the scenario in which node N2 is left discharged from a previous cycle and node N1 is precharged high. During evaluate, if inputs A0, B0, or C0 go high while node C1 remains low, then the charge on node N1 is capacitively divided between N1 and N2. In bulk technologies, it was quite possible that the resulting voltage on node N1 would droop sufficiently to cause a fail.

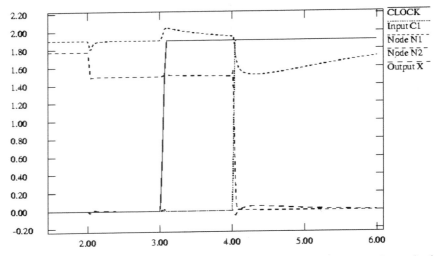

FIGURE 5.2 Simulated waveforms demonstrating bipolar effect on precharge in the circuit shown in Figure 5.1.

The severity of charge sharing is reduced with SOI, and may be illustrated with a simple example. Assume the following simple rules of thumb.

Rule of thumb: SOI source/drain capacitance is approximately 25% that of bulk CMOS.

Rule of thumb: C_{GATE} is roughly equivalent to the capacitance of 2 $C_{DIFFUSION}$, in bulk CMOS, where $C_{DIFFUSION}$ is the capacitance on a given source or drain contact.

Given input A0 goes high, node N1 had been charged high, and node N2 had been discharged low, the voltage on node N1 in Figure 5.1 may be described as

$$V_{N1}=V_{DD}(C_{N1})/(C_{N1}+C_{N2}) \tag{5.1}$$

where C_{N1} and C_{N2} are the capacitances of nodes N1 and N2 respectively.

In Bulk CMOS, the capacitance of nodes N1 and N2 may be approximated by

$$C_{N1}=5(C_J) + 2C_{GATE} \tag{5.2}$$

$$C_{N2} = 4(C_J). \tag{5.3}$$

Substituting these values into Equation 5.1 yields for **bulk CMOS**:

$$V_{N1}=V_{DD}(5(C_J) + 2C_{GATE})/(9(C_J) + 2C_{GATE}).\qquad(5.4)$$

Asserting our gate capacitance-junction capacitance equivalency rule-of-thumb produces

$$V_{N1}=V_{DD}(9/13)(C_J)$$
$$=0.69V_{DD}\text{ (normalized to junction capacitance). }(5.5)$$

Lets now recreate this calculation for the SOI equivalent circuit. Asserting our SOI junction capacitance reduction rule of thumb, we derive the voltage on the precharged node, after charge division, for the SOI equivalent:

$$V_{N1}=V_{DD}(1.25(C_J) + 2C_{GATE})/(2.25(C_J) + 2C_{GATE}).\qquad(5.6)$$

Finally, asserting our gate capacitance / junction capacitance equivalency produces in **SOI**

$$V_{N1}=V_{DD}(5.25/6.25)(C_J)$$
$$=0.84V_{DD}\text{ (normalized to junction capacitance). }(5.7)$$

Clearly, the charge-divided dynamic node voltage Equation 5.7 is improved in the SOI case over that shown in Equation 5.5 by approximately 22%. Figure 5.3 compares the simulated effects of charge sharing on dynamic precharge in both bulk CMOS and PD-SOI CMOS dominos in the 0.22μ lithography generation. In these simulations, the only modification is the substitution of the bulk device with its SOI replacement.

SOI's reduced capacitance reduces charge sharing in an additional, active manner. Referring again to Figure 5.1, node N2, the source of device 3, capacitively couples to its gate, due to gate overlap of the diffusion. As input A0 rises in potential, node N2 capacitively rises as well, further reducing the voltage difference between nodes N1 and N2 and hence the charge sharing effect. Although reduced from bulk CMOS, there still remains an observable difference in the magnitude of charge sharing when the gate of the upper device in the evaluate stack. is active compared to when it is not transitioning.

5.2.3 Miller Capacitance

Compared to the reduced source/drain junction capacitance, Miller capacitance comprising the gate overlap of the source/drain region can increase in relative magnitude. Miller Effects in evaluate devices strongly influenced performance of dynamic circuits in bulk CMOS. In PD-SOI, increased relative Miller strength emphasizes the

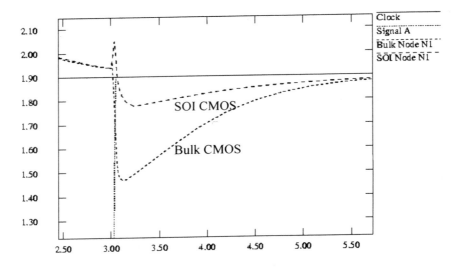

FIGURE 5.3 Charge sharing in domino circuits; bulk and SOI.

relation between junction capacitance and body voltage by exaggerating the signal "feed-through" observed.

Referring once again to Figure 5.1, as input A0 rises while C1 remained low and given a prior state that left Node N2 high, Miller Effect will couple the input signal onto node N1. As N1 rises, the charge which this device must dispose of has grown, increasing delay needed to discharge this node for a fixed device size. This additional potential is reflected in the coupling up of node N1 as shown in Figure 5.4 with input A0 rising. The bulk and SOI cases are both shown. The voltage delta caused by Miller Effect in SOI increases, as the capacitance *buffering* this coupling has been reduced in PD-SOI.

The AC behavior of an evaluate device stack in SOI comprises one of the most complex responses to characterize in SOI. The behavior of each of the components in the stack is strongly dependent on the input gate activity, prior state, output loading, as well as the device design point. Indeed, the above scenario emphasized Miller response; it is but one of a number of possible cases which might be observed, however. If Input C1 had remained high during the transition, very little Miller Effect would be observed.

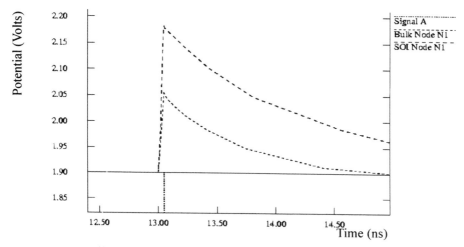

FIGURE 5.4 Miller Effect on dynamic domino summand node of circuit in Figure 5.1 on page 97, for both bulk CMOS and PD-SOI CMOS.

5.2.4 Half-Latch / Keeper Device Sizing

In bulk CMOS, the role of the "keeper" half-latch device in the dynamic domino is to replace charge lost to subthreshold and process-related leakage currents. Although charge loss on the dynamic node is directly related to the circuit's noise immunity, the keeper could not be sized large enough to provide instantaneous noise immunity. The resulting hysteresis would prohibitively penalize performance.

In PD-SOI, as discussed above, the magnitude of charge loss is potentially much greater, and for that reason, keeper sizing practices must be revisited. The performance penalty associated with increased keeper strength is still a boundary condition as it was in bulk, however. Figure 5.5 shows the impact which increased hysteresis, caused by large keeper device width/length ratios, have on the delay of the circuit. As in bulk, it is preferable to address noise problems without keeper adjustment. In addition in SOI, parasitic bipolar response is controlled better with more effective NFET evaluate tree design than by asserting stronger PFET feedback.

> **Rule of thumb:** Keeper device width/length ratio, traditionally about 5% of the effective pull-down tree width/length ratio in bulk CMOS designs, needs to be increased in PD-SOI. Approximately 25% of effective pull-down tree width/length ratio should be considered the upper limit in SOI dominos to assure reliability while still retaining performance.

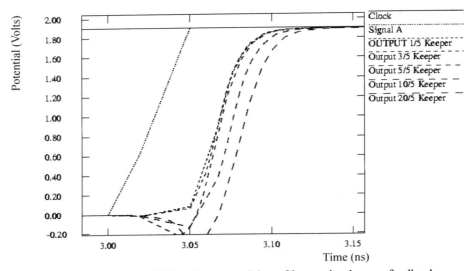

FIGURE 5.5 Effect on SOI Domino stage delay of increasing keeper feedback Width/Length ratio.

Note that sizing considerations for the keeper device should still concentrate on charge replacement rather than on providing active noise suppression.

5.3 Preferred Dynamic Design Practices

It pays to do your homework!

The use of dynamic logic, even in bulk technologies, introduced design considerations which had not been a concern in static logic [5.1]. The reader is directed to the referenced literature for a summary of fundamental dynamic circuit design considerations. Dynamic circuits carry even more vulnerability in SOI. The superior performance of dynamic logic over static logic is the motivation for identifying fail mechanisms and employing techniques to eliminate the risk. It would indeed be unfortunate to gain 30% in performance by using SOI, but then to give back 20% in performance because dynamic circuits were removed from the design!

Techniques have been offered in the literature to deal with the impacts of SOI on dynamic circuits [5.2], and are described below. These techniques, summarized in

Design Techniques for Domino Circuits in SOI

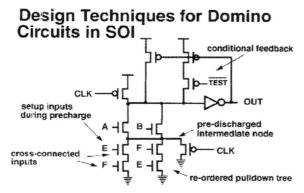

FIGURE 5.6 Dynamic Domino Design Techniques

Figure 5.6, specifically resolve common SOI design concerns impacting dynamic operation:

- Loss of precharge caused by the parasitic bipolar devices
- Loss of precharge caused by MOSFET leakage currents
- Erosion of noise immunity
- Delay variability.

5.3.1 Predischarging

In bulk CMOS, it is common practice to precharge intermediate nodes in a domino evaluation stack in order to reduce precharge loss to capacitive divider charge sharing. In SOI, intermediate node preconditioning involves very different trade-offs. There are 3 alternatives, explained using Figure 5.1:

1. **Precharge intermediate node N2 high, as in bulk CMOS**
   - This practice produces the worst noise immunity because it charges the body of devices 3, 4, and 5 high, and guarantees the worst bipolar response and device leakage.
   - The least precharge loss to charge sharing is realized when precharging intermediate node N2 high, but then with SOI, intermediate diffusions have much less capacitance (see Section 5.2.2) and charge sharing is not that big an issue.
   - The least body effect is seen, however, when the sources of device 3, 4, and 5 are high, improving performance.
   - The highest drain junction capacitance is realized when the body of devices

---

Dynamic Circuit Design Considerations   **103**

3, 4, and 5 are charged high, since the drain-body space charge region is small with the body and drain (precharge node) high. This is hard on performance but aids in noise immunity.

- The discharge performance is at its slowest, because we have intentionally inserted more charge which always must be dumped to get the evaluate tree to GND potential.

Its pretty clear this old solution now has problems in SOI.

2. **Predischarge intermediate node N2 to GND potential**
   - This practice produces the best noise immunity, as it discharges the body of devices 3, 4, and 5, and eliminates bipolar currents while reducing MOSFET leakage.
   - Precharge loss to charge sharing is now maximized when discharging intermediate nodes, but as mentioned, intermediate diffusions have much less capacitance and charge sharing was no longer a first order concern anyway.
   - The maximum body effect is seen when the source of devices 3, 4, and 5 are low, slowing the discharge of the evaluation capacitance.
   - The lowest drain junction capacitance is realized when the body of devices 3, 4, and 5 are lower, since the drain-body space charge region grows as the body and drain (precharge node) potential difference grows. This helps performance buts hurts noise immunity by removing "good capacitance."
   - The discharge performance is at its fastest, because we have eliminated as much charge as possible before evaluate even begins.

This solution sounds better, but we're really going to miss that performance. Maybe there is something in between.......

3. **Precharge intermediate node to somewhere between V_{DD} and GND.**
   - This practice achieves the desired result of reducing the opportunity for bipolar device or subthreshold MOSFET charge loss.
   - With some charge left on the intermediate node, the capacitive divider of charge to the intermediate node is negligible for most cases.
   - Body effect on devices 3, 4, and 5 is present, but a fraction of what it would have been had the node been fully discharged.
   - Junction capacitances are at an intermediate value, trading off noise immunity and performance.
   - Discharge performance is also balanced, as only a moderate amount of charge is added.

It looks like we have a good compromise! Using PFETs to discharge the node, or NFETs to charge the node assert approximately one threshold voltage worth of offset, accomplishing the results described in #3 above. Figure 5.7b below shows a preferred

SOI implementation of Figure 5.6. The designer is still warned, nonetheless, to watch

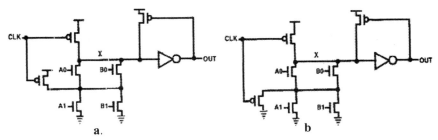

FIGURE 5.7 SOI Dynamic Domino with intermediate offset precharging. Original intermediate precharging as practiced in bulk (a); and recommended design practice in SOI (b).

out for the effects of the following on an intermediate precharged design

- Miller Feed-Forward Capacitance
- Keeper feedback strength
- Signal arrival sequence.

5.3.2 Cross-coupling of Dynamic Inputs

Aggressive domino circuits anticipate and accommodate a limited amount of pre-charge loss in bulk technologies. With care, the same practice can be continued in SOI. One means of balancing performance and noise immunity is shown in Figure 5.8 below.

As described in 5.3.1, neglect of the evaluate tree's intermediate node charge can cause precharge loss. By *splitting* evaluate devices in half, and reversing, or *cross-coupling* the order of connections in the second stack from that of the first, it is possible to limit precharge loss to half of the total possible decay while still enjoying improved performance. The approach is an alternative to the technique in 5.3.1, still achieving favorable performance while limiting charge loss.

5.3.3 Reordered Inputs

In the general case, many factors determine whether the advantage of bringing the last arriving input to the top device in an evaluate stack is the fastest in SOI as it was in bulk. What is clear, however, is that reducing the number of devices in an evaluate

---

Dynamic Circuit Design Considerations   **105**

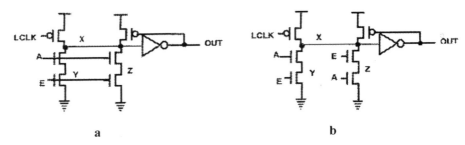

FIGURE 5.8 Cross-Coupling for balancing performance and noise. Original signal coupling as practiced in bulk (a); and recommended design practice in SOI (b).

tree of height 2 or greater which are tied to the precharge node reduces the amount of charge lost to leakage and parasitic bipolar activity. In domino stages where a common "foot-switch" provides a path to GND for a multiplicity of inputs, it is helpful to place the footswitch at the top of the stack rather than at the bottom (the bulk convention). Figure 5.9 shows a gated domino stage in which the gating input is at the top of the stack rather than at the bottom. In this manner, the total number of devices developing high body potential is reduced, and charge loss is improved.

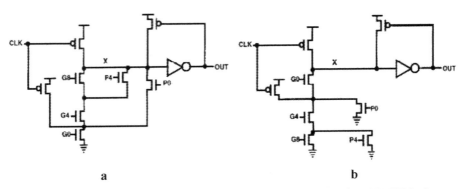

FIGURE 5.9 Reordered input sequence to mitigate bipolar and MOSFET leakage. Original signal coupling as practiced in bulk (a); and recommended design practice in SOI (b).

5.3.4 Early Setup

The concern that each of the offered remedies in this section address is how to reduce the amount of charge lost off the domino's dynamic node. The *Early Setup* technique attempts to remedy the charge loss by replacing the lost charge rather than attempting to stem it. By allowing the end of the precharge portion of the cycle to overlap the beginning of the evaluate period as shown in Figure 5.10, any charge lost to bipolar current is readily replaced by the precharge PFET device which is still on. This practice is recommended for selective use, since the overlap allows brief DC currents to pass through the chip. Since the circuit requires a customized clock duty cycle, a spe-

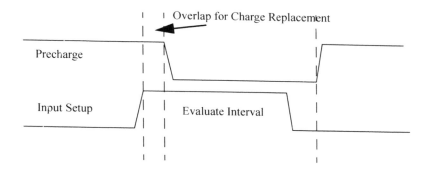

FIGURE 5.10 Timing for Early Setup resolution of Bipolar charge loss.

cialized clock regenerator must be dedicated to these books.

5.3.5 Logic Remapping

Remapping Domino Logic has been practiced in bulk CMOS for many years. By moving logical inputs out of the dynamic portion of the stage and into the static buffer, dynamic content can be limited to late-arriving critical path signals. In SOI, this technique can be used to relieve stack height; the history magnitude as well as bipolar current loss can now be limited. In Figure 5.11, a given logic function is shown before and after a portion of the dynamic evaluation is moved to the static buffer of the preceeding stage. With a device layer removed from the stack and drawn into the buffer, the resulting body voltage range of the new domino structure is substantially reduced.

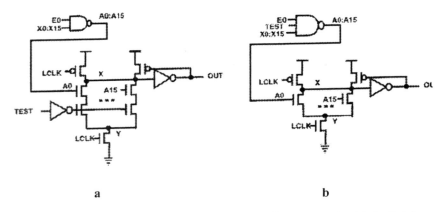

FIGURE 5.11 Logic remapping. Original signal coupling as practiced in bulk (a); and remapping into SOI to reduce dynamic stack height (b).

5.3.6 Complex Domino

Compound Domino is a well-known circuit trick practiced in bulk CMOS, which happens to be made to order for resolving SOI bipolar or MOSFET leakage concerns. The concept is simple. For extremely logically wide functions, half of the evaluate trees on the precharge node are removed to a new, second precharged node. Logic on both precharged nodes is evaluated in parallel, and both evaluated node states are then recombined in the final static buffer which produces an equivalent single bit output. In bulk, the concept was used to avoid excessive precharge or evaluate delays. In SOI, this idea is valuable because it reduces the total amount of potential bipolar current which could occur on a single node. By limiting the bipolar current spike magnitude, reasonably-sized keeper devices can then be effectively used. A dynamic domino and its complex domino equivalent is shown in Figure 5.12.

5.3.7 Dynamic Noise Suppression

As outlined in Chapter 3, the undesirable bipolar response of floating body, partially depleted SOI MOSFETs becomes more prominent at higher voltages. Although many bipolar problems may be resolved in the normal operating voltage and temperature range by the techniques just described above, additional problems arise when the chip is operated at elevated voltage and temperature. Products requiring high reliability during use undergo *Burn-In* and *Voltage Stress* prior to shipment to accelerate fail rates of parts which would have given out in the field due to processing defects. One

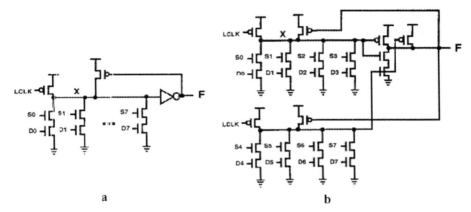

FIGURE 5.12 Complex Domino. Logically wide domino as practiced in bulk (a) incurs substantial parallel leakage; and logical width reduction in SOI which achieves same function while cutting given node's leakage in half (b).

of the most trusted means of insuring that each node in a given logic vector has seen elevated voltage is to verify that in fact the path was operational. For dynamic circuits, maintaining functionality at elevated voltage and temperature can reduce performance appreciably. Figure 5.13a shows a conventional domino circuit, and

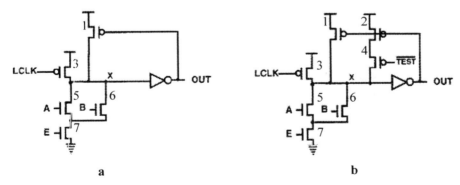

FIGURE 5.13 Dynamic Noise Suppression during high voltage stress: Common domino as practiced in bulk (a); and additional feedback in SOI to supplement charge replacement (b).

Figure 5.13b its remedied equivalent, where devices 2 and 4 have been added. Device

2 is gated, like the keeper by the output signal, but qualified by the test signal $\overline{\text{TEST}}$. During slow clock-rate test, this additional path strengthens the feedback of the path. When performance is important in the functional mode, this path is off and its hysteresis is eliminated.

A complaint with the technique in Figure 5.13 is that it adds the drain capacitance of device 4 to the evaluate node, slowing the circuit down. An alternative approach is shown in Figure 5.14. The circuit makes use of the following observation:

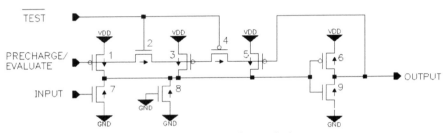

FIGURE 5.14 Dynamic Feedback re-allocation technique

1. In functional mode, precharge device 1 should be as large as possible and keeper device 5 should be as small as possible, to minimize cycle time.

2. In Burn-In mode, where the microprocessor may be operated at only 5 MHz or less, the size of precharge device 1 may be quite small, but keeper device 5 should be as large as possible to replace bipolar and MOSFET leakage precharge loss.

In stress operation, the $\overline{\text{TEST}}$ signal dynamically re-allocates a portion of the precharge device width, now segregated in device 3, over to the control of the keeper device by turning off device 2 and turning on device 4, where it will assist in noise suppression at higher voltages. The sum of the device widths of 1, 3, and 5 are the same as the width of the former precharge and keeper devices alone, avoiding added precharge/evaluate capacitance. In normal functional mode, device 2 is on and device 4 off. The effective precharge device width is maximized to reduce the total delay needed for restore.

5.4 Keeping Dynamic SOI Problems in Perspective

While it is important to address the concerns for dynamic circuits in SOI as highlighted above, the magnitude of these issues must be compared to conventional circuit design concerns already known, such as noise immunity. In this section, bipolar and charge-shared precharge loss are examined, and their impact on noise immunity is compared to that arising from coupled noise events from neighboring interconnects, as commonly battled in bulk technologies.

The following series of simulations of the dynamic 2-way AND-OR shown in Figure 5.15 assess the impact of noise when immunity is already degraded by bipolar

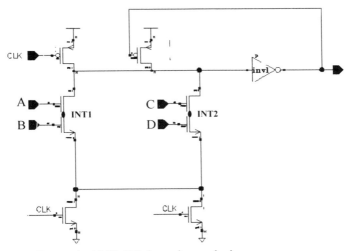

FIGURE 5.15 Dynamic AND-OR for noise analysis.

current or charge sharing [5.3].

5.4.1 Bipolar Effects and Noise in the 2W AND-OR

In the schematic shown in Figure 5.15, the intermediate diffusion between the evaluate devices is precharged to near-V_{DD}, establishing a pessimistic but not unreasonable body potential. Input A is held off, and input B is raised to V_{DD} which induces the classic SOI parasitic bipolar loss of charge from the precharge node. A noise spike caused by crosstalk on unshielded minimum-space wire is then asserted on input A, and the cumulative charge loss is measured. The scenario, which could quite likely be encountered in actual operation, is shown in Figure 5.16, and indicates that for this case, noise immunity erosion continues to be dominated by coupling events rather than the parasitic bipolar response. Surely, for more difficult settings such as that

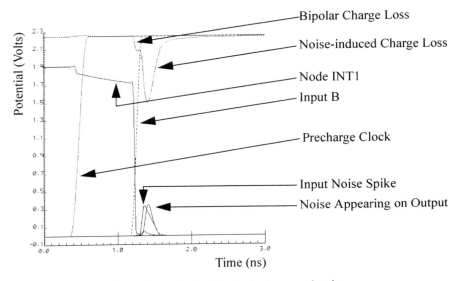

FIGURE 5.16 Precharge loss due to bipolar leakage and noise

shown in Figure 5.1, bipolar effects can be much worse, but for many circuits such as the AND-OR, conventional noise problems still are the dominant concern.

5.4.2 Charge Sharing Effects and Noise in the 2W AND-OR

Again, referring to the schematic shown in Figure 5.15, the intermediate diffusion between the evaluate devices is pre-discharged to GND, establishing a pessimistic but realistic condition. The clock is initially low, precharging node PRE. While the precharge is on, input B, which discharged INT1 goes low. Input A is then brought high, causing classic dynamic charge sharing off the precharge node. A noise spike caused by crosstalk from unshielded minimum-space wire is then asserted on input B, and the cumulative charge loss is measured. The scenario, which could quite likely be encountered in actual operation, is shown in Figure 5.17, and indicates that again, for this example and its assumptions, noise immunity erosion continues to be dominated by coupled noise events rather than charge sharing phenomena.

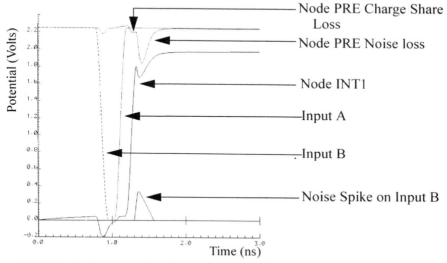

FIGURE 5.17 Precharge loss due to charge sharing and noise

5.5 Soft Errors in Dynamic Logic

The sensitivity of CMOS memory to electron-hole pair generation induced by alpha and cosmic radiation has been a known problem in bulk technologies, and has been getting progressively worse with smaller lithographies. The exposure to on-board static memory (SRAM) is associated with the ability of such an event to flip the set state of a 6-device cell [5.4]. This mechanism, and its response to SOI is discussed in more detail in Section 6.7 on page 141.

In addition to SRAM concerns, the industry expected that, with the capacitance contained in dynamic circuitry dropping with scaling, fails arising from alpha or cosmic radiation would also begin to appear in these logic structures. Bulk CMOS in the 0.18μ generation appeared to be at the edge of the cliff; PD-SOI has the potential of pushing it over the brink. Static logic has not been a concern, due to its higher fanout capacitance, and because these actively held circuits will correct temporary false states.

PD-SOI increases exposure to soft error upsets in dynamic logic in two ways. Primarily, PD-SOI reduces the capacitance on the summand node. In Figure 5.18, the capacitance on precharge node N1 comprises 3 junctions and 2 gates. With the junc-

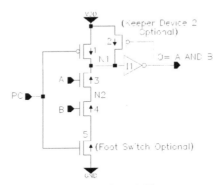

FIGURE 5.18 Dynamic AND SER vulnerability

tions abutting buried oxide in SOI, the junction capacitance drops to a fraction of its original value. The net effect is that, depending on process, the capacitance on node N1 may be half of its bulk value. Since margin in dynamic logic means precharge, reducing the summand capacitance also diminishes the critical charge loss needed to cause a fail. Charge loss beyond this Q_{CRIT} pulls the node voltage past the switch point of inverter I1 in Figure 5.18, flipping the output erroneously. Other than maintaining good precharge levels in the circuit, the only means of safeguarding against these fails is by keeping the summand node's capacitance high, to act as a charge reservoir. Increasing keeper size (Device 2 in Figure 5.18) to combat SER is fruitless, as the event occurs too quickly for the keeper to respond. It does help in keeping N1's voltage high, however.

The second mechanism specific to SOI which increases exposure to upsets is the parasitic bipolar response. Radiation injected into body of device 3 can briefly forward-bias the body-source diode, causing current I to pass from the body, and current βI to pass from the drain, tied to precharged node N1. Thus, the bipolar effect can amplify the magnitude of the SER event experienced by the circuit.

From this discussion, then, its clear that the magnitude of the radiation event can vary quite a bit. Its effect on the circuit depends critically upon:

Energy and angle of incidence of the particle
Polarity of the charge collected from the event
The role that the affected device plays in the given circuit
The structure in the given device that receives the charge

There are multiple permutations of the above factors in anticipating the effects an event can have, and it is left as an intriguing puzzle for the reader to find some of the many ways in which a radiation event can compromise the integrity of a dynamic circuit. **One clue:** consider the effect a hit would have in *any* of the devices, even on the seemingly safe static output buffer 11!

Although a concern, the probability of a soft error event in a dynamic circuit causing a fail in the machine is nonetheless remote, even SOI, for a number of reasons:

1. Dynamic Logic spends approximately half its life in reset, and is invulnerable.
2. Many circuit structures ignore inputs, and so they are don't-cares. As an example, the drains of unselected muxes, or the gates of unselected AND trees are disregarded.
3. Even if SER events occur, they may not have time left in the cycle to propagate to the next latch.
4. The dynamic circuit may already be in the state (high **or** low) that a radiation event would induce.
5. Much of the computational resource in a microprocessor is either idle, discarded, or don't-cared out, and do not participate in determining machine state. Generally the larger the chip, the lower the average activity per circuit.

Soft error upsets decrease the field reliability of a product. But like defect-related field fails, they are a fact of life for our industry. The good news is that their occurrence rate can be predicted, and the composite reliability of a total system can be budgeted to accommodate these exposures.

5.6 Dynamic Logic Performance

Hardware as well as software simulations reported in the literature for the 0.22μ technology generation indicate an average performance improvement of approximately 15-20% in moving a variety of dynamic domino logic circuits from bulk technology into SOI, with design modifications made for the technology [5.2]. The relative performance improvement of SOI over bulk, for dynamic logic, compared to other on-chip structures is shown in Figure 5.19.

When comparing the performance between bulk and SOI it is important to use delays associated with specific body potential conditions. Conventionally, performance is compared in SOI with body potentials at their steady state values.

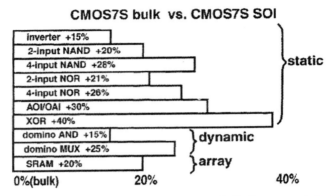

FIGURE 5.19 SOI performance improvement over bulk for a recent design [5.2]

5.7 Conclusions

Dynamic circuits have always carried additional design considerations; their mapping into PD-SOI technology requires a great deal of attention to all the possible conditions the circuit can be found in. Bulk dynamic logic design practices are clearly not appropriate in SOI, and most likely will lead to failure. Alternative design practices enable the continued use of dynamic topologies in SOI.

Retaining dynamic styles in SOI is worth it. Common domino circuits were shown to improve in performance by approximately 20% in moving to SOI, and have already been recognized to be superior in delay over static logic by an additional 20% in a given technology. It makes little sense to invest in a technology which offers performance gain, and then to revert to slower topologies which compromise that performance.

REFERENCES

[5.1]   K. Bernstein, et al., *"High Speed CMOS Design Styles,"* Kluwer Academic Publishers, 1998. ISBN 0-7923-8220-X

[5.2]   D. H. Allen, et al., "A 0.2μm 1.8V SOI 550MHz 64b PowerPC Microprocessor with Copper Interconnects," *Proceedings of 1999 IEEE International Solid State Circuits Conference*, February, 1999, pp. 438-439.

[5.3]   IBM internal analyses, Andrew Davies, IBM, Rochester, MN

[5.4]   C.Lage, D.Burnett, T.McNelly, K.Baker, A.Bormann, D.Dreier, V.Soorholtz, "Soft Error Rate and Stored Charge Requirements in Advanced High-Density SRAMs," *IEEE Technical Digest of the IEDM*, 1993, pp.821-824.

CHAPTER 6   *SRAM Cache Design Considerations*

6.1 Overview

For a number of reasons, on-board static random access memory (SRAM) is one of the most difficult functional units to map into PD-SOI technology. While the results vary with the bitline height of the array, generally speaking it is extremely difficult to realize in the SRAM the performance improvements achieved by typical logic circuits when moving to SOI. Because cache memory typically occupies a large percentage of the delay in most of a microprocessor's cycle-limiting paths, it's benefit from SOI is of particular interest.

Cache difficulties arise because many of the liabilities associated with SOI are found in structures which are key components of the SRAM. Specifically, latches, pass-gates, and dynamic circuits are integrated into the SRAM's memory cell and decoders. For this topic, the reader should first become familiar with SOI's passgate leakage in muxes, introduced in Section 4.6, "Passgate Circuit Response" on page 84.

We first examine the PD-SOI SRAM's differential write and read operations. In both operations, the height of the bitline is an important consideration. An SRAM bitline has a passgate attached to it for every wordline that crosses it. In small (short) arrays, there may be as few as 16 words on each bitline; large (tall) arrays may have 256 or 512 words on each bitline. For a majority of the discussion in this chapter, we will assume that 512 common 6-transistor SRAM cells are attached to each differential bitline pair, as shown in Figure 6.1.

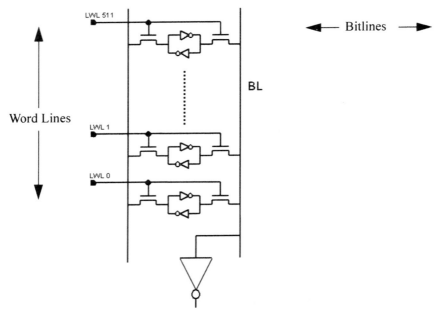

FIGURE 6.1 SRAM array cells, tied to true and complement bitlines

Simultaneous operations performed in parallel on each bitline include writing, reading, and precharging. Each require adjustment due to the SRAM response to PD-SOI, and are described in the following sections.

6.2 Writing a Cell

To write a given cell on the bitline, The write enable signal WR goes high, and *either* the true *or* the complement bitline is pulled down by the large write inverter after the bitlines have been precharged high as shown in Figure 6.2. To write a logical 1, the selected cell's passgates to the true and complement bitlines are opened by raising it's

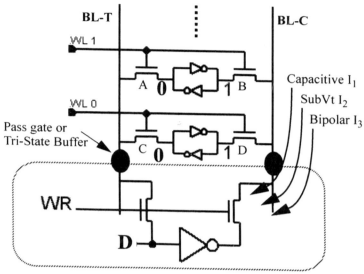

FIGURE 6.2 Write Circuit on SRAM bitlines

selected word line, say WL 0, high. With input D a logical 1, the true bitline's write inverter holds the true bitline high, the complement bitline write inverter pulls the complement bitline to GND, and the cell differentially flips to the desired polarity. The minimum delay needed to reliably write data into a cell depends upon the conditions of the other 511 cells on the bitline. The slowest write is when each of the other 511 cells contains a logical 0, and has been sitting at that state long enough for the devices comprising those cells to reach steady state body bias. The bitline write driver must then sink currents from a number of different origins. SRAM designs usually require that the bitline be fully discharged to GND, meaning that the write inverter must sink the entire range of possible currents caused by the SOI device. Three independent sources of current are described below.

6.2.1 Capacitive Discharge Current

Initially, the capacitance of the bitline wire, its parasitics, and the 511 pass gate source diffusions/junctions attached to it, as shown in Figure 6.1, begins to discharge through the appropriate write inverter. This is represented as current I_1. Because the bodies of the pass gates on the true and complement sides of the selected bitline have converged on different steady state voltages based on the data content of the cells, the resulting total capacitance on those bitlines and its capacitive discharge rate can assume a wide range of values. Even though, in SOI, the area component of the pass gate diffusion capacitance is removed from the source junction, the capacitance between the source and the body can be very high. Over an extended period, both sides of the junction will converge on the same potential and the resulting depletion width will be very small. This small width results in an unexpectedly-high capacitance on the given bit-line and increases the bitline discharge delay. Figure 6.3 implicitly affirms this bitline capacitance delta. A delay difference is observed when reading a zero onto a bitline in which all the other bits are ones, compared to a bitline in which all the other bits are zeros. With a sufficiently large bitline inverter, this has not proven to be a problem for

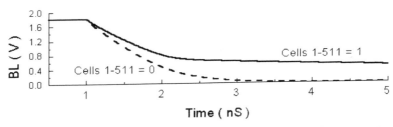

FIGURE 6.3 Bitline Voltage during a read operation for a 512 cell bitline, with all cells containing a 0, and containing a 1. Final discharged level is dependent on bitline capacitance presented by cell state [6.1].

functionality, but it does introduce variability in the amount of bitline write-time needed.

A pair of bitlines rarely have equal capacitance; therefore it is prudent to anticipate the scenarios that differential capacitance can create.

6.2.2 CMOS Subthreshold Leakage

Next, conventional CMOS subthreshold leakage current I_2 from up to 511 cells through their respective pass gates can leak charge originating from the PFET on the high side of the cell to their bitline. Referring to Figure 6.2, assume 511 cells such as the cells shown all contain "0"s. Since the bodies of all the pass gates tied to the com-pliment bitline, represented by devices B and D, are charged to V_{DD}, their threshold voltages are at their lowest value. This can produce leakage currents that are more than an order of magnitude larger than comparable passgates in bulk. This leakage begins as soon as the complement bitline begins to fall, and interferes with the devel-opment of voltage differential on the bitlines. Although the worst case for a given bit-line occurs when trying to pull down the high-side bitline of 511 of 512 cells which have been flipped in that direction for an extended period of time, the subthreshold current arising from even half that many can still present a problem.

6.2.3 Diode and Bipolar Currents

Finally, body discharge current ("I") and the resulting bipolar current ("βI") from each of the 511 cells, (cumulatively I_3) appears as soon as the bitline drops below one diode voltage under V_{DD}. The bodies of pass gates B and D in Figure 6.2 had been at V_{DD}, and so, as the source side of these passgates (a.k.a. bitline) for that wordline are pulled low, the body-source diode forward-biases. Charge spills from the bodies of pass gates B and D to their source diffusions tied to bitline BL-C, and causes a bipolar current pulse which lasts until the body charge is depleted. In the meantime, the write inverter must be equipped to handle this temporary current in a timely manner.

The SRAM arrays in large system engines often "dot-OR" identically addressed bit-lines from a number of "ways" or separate subarray leaves to the same write pull-down circuit. The resulting write load presented to the write inverter in PD-SOI can be staggering. Although each "way" is usually gated, the currents to be sunk are multiplied by many times.

6.3 Reading a Cell

Large SRAMs typically read the differential bitlines using a cross-coupled, strobed sense amplifier attached to the true and complement bitlines. The sense amplifier, shown in Figure 6.4, is a high gain comparator which samples the voltage developed between the bitlines after a precise delay interval. The differential between the bitline and the complementary bitline ranges from 5-25% of V_{DD} depending on the aggressiveness of the design and the speed needed in the array. With the bitline discharge limited to only 25% below V_{DD}, little bipolar current is present, as the voltage on the bitline barely forward-biases the diode between the source and the body of the pass-gate. The increased capacitance across this junction must still be considered.

> **Rule of thumb:** Parasitic bipolar currents only need to be considered if the source of the passgate transistor is pulled more than a diode drop beneath the body.

Two issues immediately surround reading of a bit cell. Reading the cell requires that sufficient differential be developed given the highest possible *leakage of unselected cells*. One must also insure that sufficient differential voltage is developed to cover the maximum *bias which could accumulate in the sense amp*. Finally, *timing* must include delay variation in other support structures. These issues are addressed below.

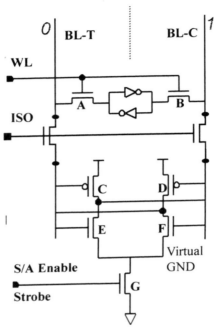

FIGURE 6.4 SRAM's Differential, cross-coupled sense amplifier, used in reading.

6.3.1 Read Leakage

The leakage mechanism challenging the development of read differential, MOSFET subthreshold current through the off cell's pass gates, is identical to that occurring during the write operation and described in Section 6.2.2. In the read operation, this leakage remains as DC current because the bitline is usually not discharged low enough to empty the charge in the bodies of devices A and B in Figure 6.4 on page 124. The magnitude of the leakage current is significantly smaller than for the write scenario since the potential across the passgate device is now only a fraction of V_{DD}.

6.3.2 Sense Amplifier Bias

Ideally, the sense amplifier of an array column contains closely matched devices to improve the ability to discriminate small voltage differentials across the true and complement bitlines[1]. With the body on the sense amps electrically floating, even identical devices will become mismatched.

Body Charge-Induced Bias

If a sense amp has read a logical 0, it will more easily read a logical 0 the next time. If a logical 0 is read many times in a row, the sense amp will have built up a preferential bias toward reading a logical 0. Referring to Figure 6.4, if the complement bitline has remained high and the true bitline repeatedly discharges and precharges, the body potential of device E will be lower, which **increases** its threshold voltage and **increases** the switchpoint voltage at which the true side inverter will switch which makes it easier to read a logical 0 again. This *bias* is enhanced by the opposite side of the latch: body potential of device F is higher. This higher body voltage will **decrease** the threshold voltage of NFET F and **decrease** the switchpoint of the inverter on the complement side, making it easier to hold the logical 1 value on the compliment side.

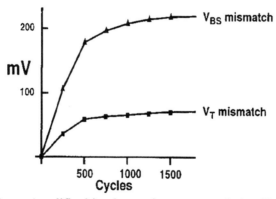

FIGURE 6.5 Sense Amplifier bias due to charge accumulation [6.2].

Let's look further at the transitions that lead to device F having a lower threshold voltage than device E. When the sense amp has latched the true bitline at a logical 0 and the compliment bitline at a logical 1, the body of device E is higher than the body of device F, since device E has $V_{DS} = V_{DD}$ and device F has $V_{DS} = 0V$ and equal to GND. However, when the sense amp returns to its precharge state with the sense

---

1. Layout practices which ensure close device matching are known in the art, and include (a) using identical gate and diffusion polygons, (b) keeping gates close to one another, and (c) keeping gates oriented in the same direction.

enable at GND, device E only has its source precharging, while device F has both its source and drain precharging. This elevates device F's body to a higher potential than device E. If the sense amp is not strobed for a long period of time, the potential of the body of devices E and F will eventually equalize. Thus, the worst case mismatch for a sense amp is when it spends the largest percentage of its time holding the latched read data and is run at its highest frequency.

Again, when it is time to read a logical 1 on the true bitline, the bitline differential must be large enough to overcome this preferential mismatch within the sense amp. This will add delay, as the sense amplifier's bitline sampling interval must be delayed until adequate bitline differential is developed. The worst case steady state in the array is when a logical 0 has been read for thousands of cycles before a logical 1 needs to be read, or when a logical 1 has been read for thousands of cycles before a logical 0 needs to be read. Again, these two worst case bias scenarios arise from the accumulation of charge in the bodies of transistors in the sense amplifier on the high side.

Body Contacts

One means of reducing sense amp mismatch and bias is to electrically connect the bodies of critical devices to a known value. Two alternatives are described. In one approach, the bodies of devices C, D, E, and F in Figure 6.4 can be tied to their respective sources using the technique shown in Figure 6.6. or to power supplies (to

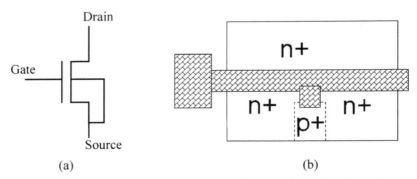

(a)   (b)

FIGURE 6.6 Body connections to reduce bias effects. Schematic (a) and layout (b) using a source-to-body contact are shown. Images are not drawn to correct proportions.

mimic bulk CMOS) or other external interconnects using the technique shown in Figure 6.7. The resulting schematics of the sense amplifiers are shown in Figure 6.8

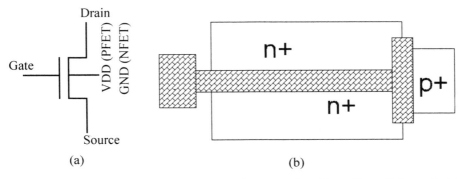

FIGURE 6.7 Rail sense amp body connections to reduce bias effects. Schematic (a) and layout (b) using a rail-tied body contacts are shown. Images are not drawn to correct proportions.

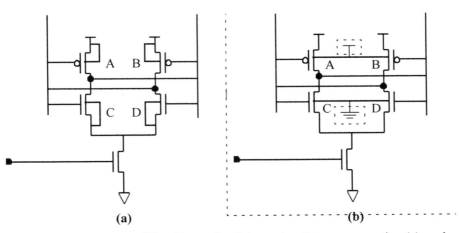

FIGURE 6.8 Sense Amplifier Alternative Schematics; Source connection (a), and body-body connection (b).

While mimicking bulk CMOS returns the designer to familiar territory, it guarantees the longest evaluate times. Cross-tied bodies can simply share accumulated body potential, and the rail connections shown in the dashed boxes in Figure 6.8 may be omitted. The sense amp NFET devices C and D, for example, will have *variable* but *higher* body potentials and lower threshold voltages than if their bodies had each been tied to GND, forcing them to *fixed* but *permanently lower* body potentials and higher collective threshold voltages. Figure 6.9 shows the delay caused by each scenario.

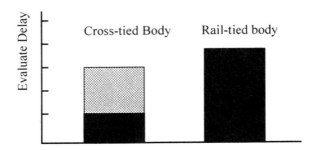

FIGURE 6.9 Timings under cross-tied and rail-tied sense amp scenarios, as shown in Figure 6.8b. The solid region indicates the minimum expected delay: the cross-hatched region shows the delay range associated with the floating body. Note that even in the worst case, cross-tied body sense delay is faster than rail-tied body sense delay.

Since existing body contact technology is not elegant and adds capacitance to the sense amp, tying bodies together or to the supply rail can actually add more delay. Figure 6.10 shows the resistance between body contacts for a MOSFETs with varying device lengths [6.3]. From the plot it becomes apparent that a large, finite RC delay constant must be anticipated when attempting to force body bias in an AC coupled body.

It sometimes is more economical to avoid the body contacts and instead wait for a larger bitline differential to be developed before enabling the sense amp by raising its strobe input. For-high-reliability grade parts however, it is essential that in test, the pattern set used be economical and effective. This suggests that although it may cost performance, body contacts will reduce the probability of defects from escaping. Each SOI technology must be evaluated for the fastest safe method for reading data after reading the opposite data for many cycles.

6.3.3 Read Sense-Amp Strobe Timing vs. Read Frequency

To determine the correct time to sample or *strobe* the sense amp, an evaluate delay line is commonly employed. This delay line is initiated by the same clock which enables the word lines in the array each cycle, and is used to tell the sense amplifier

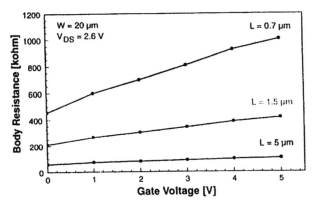

FIGURE 6.10 Measured resistance between body contacts for a MOSFET of varying gate lengths.

when it is safe to sample the developed differential across the true and complement bitlines. This delay circuit is active every cycle and so reaches a stable SOI steady state delay from cycle to cycle. The wordline, on the other hand, only goes high when it is selected, and may be as slow as the first switch time, as fast as second switch, or has also achieved steady state delay. In addition, the bitline may have been discharged and recharged, or may have remained high. The activity factor of given address is determined purely by the input instruction stream, and the data in that address is determined purely by the data stream. Therefore, a substantial offset may exist between the delay line and the bitline.

> **Rule of thumb:** The potential delay variation caused by sense amp mismatch usually dominates the determination of sense amp delay much more that than the delay difference caused by the lack of tracking between read history and delay line history.

At the very least, however, it is clear that their performance does not necessarily track with each other. This delay line needs to be timed so that the sense amplifier does not strobe the bitline before sufficient signal differential has been developed.

An alternative available to the designer is to tie the bodies of all devices (except the array devices themselves) in order to reduce the variability. This option inevitably introduces additional area and capacitance, and usually is not productive.

6.3.4 Single Polarity SRAMS

Small arrays are occasionally designed with a single ended cell. In these designs, there is no sense amp, but instead a simple inverter at the end of the bitline. This requires that the bitline be fully discharged to switch the inverter safely. Under this scenario, the bipolar leakage and the increased capacitance limit the speed of the bitline pulldown and must be included when determining best case and worst case delay.

6.3.5 Word Line History

Clearly, the timing of the sample clock to the SRAM's sense amplifier is critical to the functionality of the array. And above, we have examined the temporal variability in the development of bitline differential caused by history effects in the bit cell and in the sense amplifier. Our treatment of the SOI SRAM read operation is incomplete however, without examining the timing variability of the other structures which participate in the critical path through the SRAM.

The same clock which initiates the delay line ultimately ending the sense amplifier sample period also enables the selected wordline to be driven high across the array, coupling the true and complement sides of each cell in the word to their associated bitlines. This path is shown in Figure 6.11. It follows, then that variation in the time it

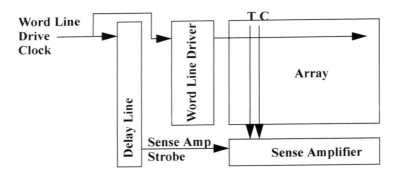

FIGURE 6.11 Word line decode timing path in array.

takes to drive the selected word line high will contribute to the variability in total sample time. Variation in word line delay indeed exists in SOI, and arises from history effects in the word line decoder and driver. If a given address has been read repeatedly for thousands of cycles, the body voltage of its word line driver devices will rise, their threshold voltages will decrease, and the resulting delay will be reduced. Conversely, a word that has not been read for a millisecond or more will have driver body voltages which are near their lowest potential, causing the greatest delay. Although one might presume that the probability of reading the same word repeatedly decreases with increased array size, specific operations within the microprocessors can quickly push the array into a worst-case word line history scenario. Indeed, decoder history must be included in the total read delay budget.

6.4 Cell Stability and Cell Bias

Now that the behavior of the SOI cell during fundamental array operations has been established, we need to update the practices used in designing a robust memory. The foremost issue, establishing cell read stability, is addressed first. Prudent cell device size selection is key to achieving performance and cell stability simultaneously.

The ability of an SRAM cell to retain data most recently written into it while being read is referred to as *cell stability*. The corruption of the cell's contents while being read is known as a *read disturb*. The designer must select device sizes which assure a minimum signal differential voltage will be produced on the precharged bitlines in a fixed delay, anticipating a plethora of influences which can introduce bias into the memory cell and diminish bitline differential voltage. After reviewing the concept of transfer ratio, several sources of bias in PD-SOI will be explained.

6.4.1 Transfer Ratio

SRAM *Transfer Ratio* is the ratio of the width/lengths of the pulldown device to the transfer device; referring to Figure 6.12,

$$R_{TRANSFER} = W/L_{T3}/W/L_{T5} \tag{2.1}$$

Cell stability during the read operation is determined by the transfer ratio built into the cell. Depending on topology, the relative strengths of the pulldown devices 3 and 4 to pass gate devices 5 and 6 can vary due to their floating bodies, causing the transfer ratio to assume a wider range of values. The ratio of pass gate to pulldown is a critical SRAM design parameter.[2] If the pass gate becomes too strong with respect to the pull down NFET in series, the cell being read may be overwhelmed by bitline charge and flip to a false data state. If the pass gate becomes too weak with respect to the pulldown, the cell may not develop sufficient differential in the time allocated[3].

---

2. Typical transfer ratios range from 1.8 - 2.0 in bulk, 1.8 - 2.8 and SOI SRAM cells. The wider range is due to the uncertainty asserted by the floating bodies.

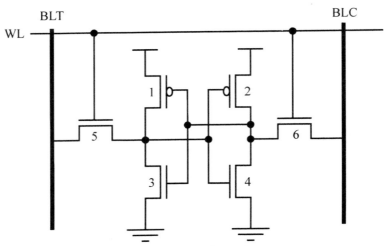

FIGURE 6.12 6-Device SRAM Cell Transfer Ratio

6.4.2 Mistracking

The prudent selection of device sizes in the 6-device cell provides sufficient preservation of signal differential in the cell while executing high speed reads, as well as the ability to write the cell in a timely manner, in an economically sized cell. The SRAM cell is vulnerable to bias just as the sense amplifier previously described is, but from different sources. The cell design must allow sufficient voltage differential which potentially could be consumed by the following mechanisms:

1. PFET to NFET Offsets

    Variation in the relative strengths of PFET and NFET transistors comprising the cell, embodied as differences in threshold tracking from NFET to PFET, can assert bias in the cell by altering the effective beta-ratio on either side of the cell. This imbalance causes differing pull-up strengths across the cell, resulting in shifts of switch point and a reduction in noise margin.

---

3. A potentially powerful fix to many of the SRAM's PD-SOI problems is blocked from use by the transfer ratio conundrum. If each of the bitlines were precharged to $V_{DD}/2$ rather than V_{DD}, excursion in body voltages would be *dramatically* reduced. Unfortunately, with the bitline starting at $V_{DD}/2$, the cell during a read is moved substantially *closer* to switching, as both sides hover near the cell's switch point.

---

2. NFET-NFET and PFET-PFET Offsets

A well-characterized process has known tolerance in the variability of device width, length, and threshold voltage between two identical transistors. This variation has both systematic and random components, and comes from both mask and process variability. These distributions should be well known and made available to the circuit designer. Their influence on cell stability can not be understated. All three parameters influence the strength of an individual device, and therefore can impact the overall cell balance. As stated in Section 6.4.1, transfer ratio must be precisely controlled, and these three parameters can assert first-order effects.

6.4.3 Self Heating

A characteristic of an effective SRAM cell design is high device density. It was pointed out in Section 3.6.1, however, that devices in close proximity to one another which share common diffusion islands have the ability to transfer heat to one another, and affect one another's transconductance. Further, typical SRAM cells, which share bitline contacts to conserve space, place the pass-gates of adjacent cells at the minimum diffusion-contacted gate pitch spacing, as shown in Figure 6.14. The pass gates are sized to achieve a fixed cell transfer ratio, which clearly can be disturbed by heat transferred across the common diffusion island they share.

Self-heating in the SRAM cell is an easily-realized mechanism. Sequential reads of a given word causes the same cell to discharge the same side of the bitline repeatedly through the pass gate and pull-down device. At high frequencies, the duty factor of the two devices is high, and results in a gradual degradation in transconductance as they heat up. The amount of self-heating is a function of the load the device is driving, and is depicted graphically in Figure 6.13. This degradation does not usually strongly affect that particular cell's stability, as lower pass gate strength tends to stabilize the 6-device cell, but has the ability, incidentally, to affect necessary read delay. The adjacent pass gate, belonging to the next cell however, also becomes degraded in transconductance, but the data it is reading may be high *or* low, and is independent of its neighbor's contents. This additional bias should be modeled and anticipated.

6.4.4 Body Bias

Just as the sense amplifier accumulated bias by sitting in one state until the body voltage in its transistors reached equilibrium, SRAM cells can also suffer the same fate. The difference is that body ties to rail, added to sense amplifier devices to reduce bias, can not be afforded in the array because of the area and performance penalty. As a

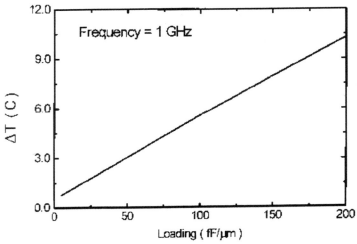

FIGURE 6.13 Self-heating dependence on load. SRAM cells can experience relatively high bitline loading, depending on logical height [6.1].

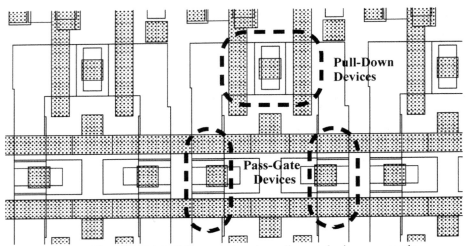

FIGURE 6.14 Test SRAM Cell Layout showing adjacent devices exposed to transient heating after repeated reads (Compliments of IBM Microelectronics)

result, the design must accommodate full possible SRAM cell bias. Although bias may accumulate in the pull-up, pull down, and pass gate devices, the body charge of the pass gate device has the most influence on behavior, as it's position in the cell's schematic allows the greatest swing in body bias. That is, with the capability of its source to reach the supply rail, much more body-induced variability may be realized.

6.4.5 Supply Rail Droop

Power supply distribution is a common cause of read problems in bulk CMOS due to the reduced overdrive it causes, and due to a mismatch in the voltage of "1" outputs. With lower overdrive, other small variations asserting bias into the cell command a larger percentage of the total differential and can induce errors more easily. As a result, the differential which the cell is able to provide to the sense amplifier is diminished, either allowing less sense margin, or asserting a greater built-in bias which needs to be overcome.

In SOI, this mechanism persists, but may be somewhat complicated by the floating body. While it remains true that bias sensitivity increases inversely with voltage, it is also noted that junction leakage diminishes as well with a lower supply, providing a measure of compensation. The net contribution of body potential to cell bias is dependent, then, on voltage and temperature, as well as on the ratio of forward-biased to reverse-biased junction diode currents as previously discussed.

6.4.6 Body-to-Body Coupling

In Section 6.4.3, it was noted that adjacent devices can influence each other due to their power dissipation. In addition, adjacent pass gate devices also may accumulate bias from one another via the parasitic JFET or PFET phenomenon described in Section 3.7.1 on page 60. This results in a neighboring device unexpectedly exhibiting substantially lower thresholds than anticipated by it's recent use history, due to the added body charge it picked up from its neighbor's body. The pass gate's lowered threshold voltage introduces cell bias which can lead to a read disturb. The read disturb arises from excessive *differential* coupling of the true and complement bitline precharge into the cell through the given cell's pass gates during a read, which turns on the opposite side's pull down NFET, causing a flip of the cell data content. The pass gates, containing differing charge due to the body coupling, create this differential.

6.4.7 Defect-Induced Bias

SOI SRAM read operations are especially sensitive to defects which were inconsequential in the bulk CMOS predecessor. In Figure 6.4, reading was accomplished by raising the selected word line high, coupling the selected cell to both BL-T and BL-C through devices A and B. Cell device sizes have been selected to achieve a switch point which requires differential signals on BL-T *and* BL-C to accomplish a write. If both bitlines are high, however, reading may be safely achieved without flipping the cell, as the differential within the cell is not overwhelmed. As described on page 125, the SOI array designer needs to verify this condition for all possible body potentials.

Insufficient bias margin leaves the design vulnerable to **defect-induced bias**. This bias may be a result of leakages resulting from poor processing. In bulk-CMOS with its substrate tied firmly to GND, these defects were often too small to be detected and screened during DC testing, and had no effect on the cell. In SOI, these defects may be no larger in magnitude than they were previously, but the floating body becomes much more sensitive to their effects. These defects can establish small built-in differentials in the cell, which when added to bias caused by use history, can induce a fail. **As a result, testing and screening for defects must be performed in AC mode in SOI, and often at functional speeds**[4].

Examples of *shorting defects* include silicide gate-source and gate-drain paths, gate oxide currents from gate to body, and problems arising from the formation of the contact between diffusions and interconnect. *Opens defects* include highly resistive contacts to the supply rail or GND. The reader is encouraged to consult the appropriate literature for more detail on the electrical properties of physical defects [6.7].

Defect-induced bias can only be exposed by rigorous testing, using array data patterns which will produce the extremes in history for each cell in the device under test.

6.4.8 Passive History Effects

History effects occurring in the array are no different from those exhibited in other logic structures. *Passive History Effect* occurs in array devices when only the drain or source of a device moves, without the gate being active at all. For example, if WL1 shown in Figure 6.2 has been inactive for an extended period, and the cell contents have not changed, then both the gate and the source of devices A and B are quiet. Yet these devices will exhibit history, based on the charge induced in their bodies by the movement of the drain only. Over extended periods, the history induced by drain-only activity can produce body voltages in the same range as an active device. If the source is high, the body can reach potential V_{DD}, and similarly if the source is low, GND potential can eventually be observed in the body, both given appropriate drain movement.

---

4. Often referred to as "at speed" test.

6.4.9 Making Sense of All This Bias!

From the above sections, there is clearly an array of mechanisms which have the ability to assert bias into a given cell design. In the worst case, each of them contributes imbalance to the differential structure *to the same preferred state*. The question this begs then, is how to assess whether a given SOI cell design has sufficient differential voltage margin to withstand the cumulative adverse influence. The presence of the floating body drives the designer to use a variation of the bulk technique to verify cell robustness

1. Size all transistors to the shortest possible lengths likely to be encountered over the life of the product, including across-chip length variation (See reference) [6.6].

2. In the simulator, assert the worst possible device width, length and threshold mistracking magnitude on the SRAM cell devices in such a way that it maximizes differential switch bias. In the same manner, diminish the PFET transconductance by the maximum called for in the process description (see Section 6.4.2).

3. Initialize the bodies of the cell transistors to the worst achievable potentials which also exaggerate the cell bias (see Section 6.4.4).

4. If the model allows, assert a temperature difference which reflects high read utilization immediately preceding the simulated switching (see Section 6.4.3).

5. Establish the product's worst application condition for stability via hardware characterization of modeling, i.e. Burn-in.

6. Assert any defects that are known to be systematic and unavoidable, i.e. defects that are expected to appear on every chip (See Section 6.4.7).

7. In the simulation, prepare waveforms for the sense amplifier strobe clock and word line drivers which contain the maximum expected clock skew and use history to maximize the exposure to timing changes caused by actions 1-5 above.

8. Incrementally simulate to identify the minimum bitline differential required to avoid a fail. Figures of merit may be the minimum multiplier of pull down or pass gate device width causing a fail, or the magnitude of additional body bias which causes a fail. Any number of measures are acceptable.

It's no surprise that the margin finally selected for the 6-device cell requires a trade-off in reliability, performance, and area.

6.5 SRAM Noise Considerations

An SRAM array at full speed is not exactly a quiet place.

The conventional noise concerns monitored by bulk CMOS SRAM circuit designers are still very important. Referring to Figure 6.15, there are additional SOI mechanisms which must be watched.

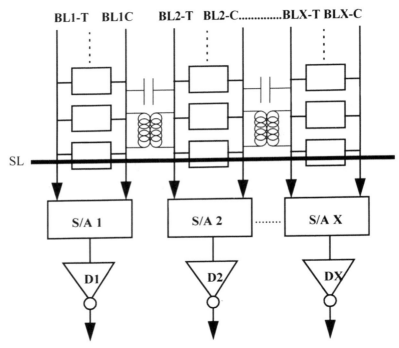

FIGURE 6.15 Noise considerations in SOI SRAM cells

Classic Lateral Coupling

Classic lateral coupling of one cell's moving bitline can inductively or capacitively assert a false signal which is detected by the sense amplifier of a neighboring bitline. Referring to Figure 6.15, bitline BL1-C can induce a false signal on the compliment bitline BL2-T, upsetting its respective sensed output. In SOI, this problem is of great concern, with the reduction of the source and drain capacitance, the total capacitance on a bitline may be less such that the capacitive coupling will reduce the bitline differential more than in bulk. As mentioned before, the body preconditioning can lower the noise immunity of the associated sense amplifier tied to the BL2 differential pair and result in a more easily flipped sense amp.

Differential Bitline Capacitance

Because the true/complement bitlines are not likely to have the same capacitance due to differences in body potential, a common-mode noise event (say a switching line SL in Figure 6.15 crossing both bitlines) can assert a differential noise voltage which will appear at the sense amplifier S/A 1. The same noise source, formerly common mode noise in bulk CMOS, now couples to different capacitance values on each bitline, creating a differential noise magnitude on each.

Pass Gate Read Leakage

Leakage during the read operation can subdue the development of differential signal provided to the sense amplifier. In the worst case, if the lines BL1-T and BL1-C droop too low, the levels approach the unity gain point of the drivers outputting the sense amp results, eroding noise margin.

Common Mode Supply Rail

The elimination of the N-Well implanted in bulk hardware to establish PFETs in p-substrate wafers also removed a native source of decoupling capacitance. In Figure 6.16, the N-well providing the PFET its body, usually tied to V_{DD}, formed a junction with its associated capacitance directly to the p-substrate, tied to GND.

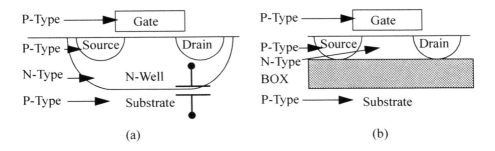

(a)   (b)

FIGURE 6.16 Loss of natural N-Well decoupling capacitance in SOI. Natural decoupling in Bulk CMOS PFET's NWell (a); SOI PFET (b).

Because each SRAM cell on chip contained 2 of these structures, substantial natural supply rail decoupling was provided by large SRAM arrays. In SOI, this is gone, and if it is determined that additional decoupling needs to be added, the designer must create separate capacitance structures, using the given technology's layout rules with accumulation mode device gate oxide capacitors.

6.6 Precharging Circuitry

The bitline precharge before reading or writing the array potentially requires greater delay in PD-SOI due to the greater range of capacitance each bitline may assume. The variability in charging delay is associated with the change in the pass gate's drain-body capacitance at different body potentials, as described in "Capacitive Discharge Current" on page 121. The conventional schematic for recharging SRAM array bitlines, shown in Figure 6.17, is essentially unchanged from the bulk CMOS approach. Body contacts are recommended, as shown, but are not essential. The body contacts

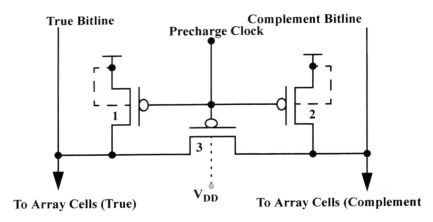

FIGURE 6.17 Bitline precharge circuitry, with optional body ties.

assure uniform precharge delay via PFET devices 1 and 2, and eliminate history effects interfering with the equalization of line potentials in PFET device 3

6.7 Soft Error Upsets

Soft errors cause hard problems.

The vulnerability of the static 6-device SRAM cell (Figure 6.12) to alpha and cosmic radiation incidence events has become profound with continued process scaling [6.4]. Once only a concern in the DRAM domain, the reduction in capacitances with successively improving lithography generations has raised the SER[5] concern in SRAM memory and in selected logic topologies as well. If not addressed through other means, higher reliability requirements often force the SRAM designer to increase SRAM storage cell size by using non-minimum layout dimensions [6.5]. SOI's superior innate resilience to incident alpha particles and cosmic rays again makes this technology timely.

Soft errors refer to false data states induced in the microprocessor logic or memory by the instantaneous introduction of high amounts of unexpected charge, created by a radiation event. Charge can be created as a result of incident alpha particles or high energy protons or neutrons (cosmic rays). The reader is directed to more thorough treatments of this topic in the literature [6.8].

The SOI MOSFET's response to soft error events is a balance of a number of compensating effects. Assume an incident alpha particle strikes the left side pulldown NFET in Figure 6.18 above. The event can deliver charge and energy to any of the structures within that NFET as shown in Figure 6.18. The result is the balance of competing effects.

1. Radiation events cause less charge generation and collection in SOI, as the sensitive region through which the alpha particle passes is now limited to the active silicon layer above the SIMOX (or BOX). The thickness of epitaxial silicon in the bulk device provides a larger region for electron-hole pair generation. The particle creates ionized charge along its trajectory through the device, which is quickly swept up by the reverse-biased junctions.

---

5. Soft Error Rate, the fail rate of an array due to alpha and cosmic radiation, commonly measured in fails per thousand hours per thousand bits.

SRAM Cache Design Considerations   **141**

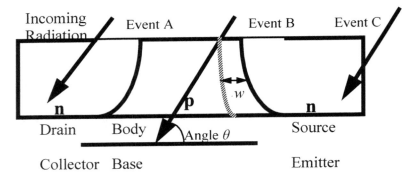

FIGURE 6.18 Radiation striking SOI NFET structures, noting MOSFET/Bipolar duality

The amount of charge generated in SOI is strongly dependent upon a number of factors, all of which have a probability distribution associated with them. These include

the energy of the incident particle,

the vertical angle of incidence with respect to the surface,

the horizontal angle of incidence with respect to the device orientation,

the channel length of the device,

the location of incidence with respect to the device, and

the local impurity concentration.

2. *Both* diffusions are now vulnerable, however. In bulk CMOS the source, tied to GND potential is quiet, and strikes to the source are fairly inert. SOI's parasitic bipolar device, however, multiplies a strike to the drain, as

$$I_D = (\beta+1)I_{minority} \quad \text{(drain event A)} \qquad (2.1)$$

or to the source, as

$$I_D = \beta I_{minority} \quad \text{(source event C)} \qquad (2.2)$$

where β is the gain of the bipolar device.

3. Q_{CRIT} for minority carrier injection, not including bipolar action, increases with operating voltage, making the cell harder to upset at higher V_{DD}. This is true of SOI as well as bulk. The dependence on voltage is stronger (less favorable) for the bulk device, as shown in Figure 6.20. Despite SOI's intrinsic SER advantages

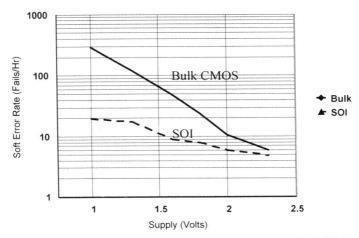

FIGURE 6.19 Soft error rate for identical array test hardware built in bulk CMOS and in PD-SOI, when placed under an active alpha particle radiation source.

over bulk, continued scaling will further reduce Qcrit[6] and will remain sensitive to voltage, as shown in Figure 6.20.

> **Rule of thumb:** Soft error upsets become more likely at lower voltages, both in SOI and in bulk. SOI just doesn't get as bad as quickly. At nominal voltages, SOI SER of an SRAM cell may have a 2X advantage over the SER of a bulk cell with identical dimensions. At reduced voltages for a given partially-depleted technology, that advantage may increase to 20X.

4. The *Early Effect*, however, must be monitored. At higher voltages, junction depletion width w grows, reducing the effective CMOS electrical channel length, and unfortunately the parasitic bipolar device's effective base width. This voltage dependent base width modulation can increase bipolar gain and vulnerability to SER.

5. Unless tied to a supply rail, the body potential of SRAM transistors are likely to be greater than zero. This elevated body voltage can differentially reduce the threshold voltage of the bistable circuit, reducing Qcrit and threatening cell stability. The area penalty associated with body ties precludes them from use in most array settings

---

6. Qcrit refers to the critical amount of charge which must be injected in order to induce a bit cell fail. Qcrit is measured in femto-Coulombs

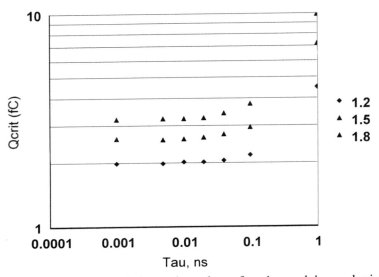

FIGURE 6.20 Typical critical charge dependence for advanced deep-submicron SOI technologies, based on modeling by the authors.

6. SOI junction capacitance reduction removes stabilizing ("good") capacitance in the SRAM cell. Because the cell response is a capacitive divider of the injected charge between junctions and body, the lower capacitance makes upset events somewhat more likely.

The net is that for a given technology, soft error rate may be improved by approximately 2X at nominal V_{DD} by moving a MOSFET structure from bulk to SOI. Although at lower voltages SER in general increases, the SOI SER advantage over bulk becomes substantial, depending on device design point. The reader is reminded that a technology's radiation event response is **highly** variable, and has profound design and process sensitivity, specific to the particular instance.

While we're on the topic, it is noteworthy that, due to scaling, dynamic logic on board the microprocessor may suffer from soft error upsets as well. This topic was addressed in more detail in Chapter 5.

6.8 Array Test in PD-SOI

If there is good news, it is that the nature of array test lends itself nicely to uncovering and diagnosing problems. Specially-crafted patterns can easily create the worst case electrical conditions needed to expose the above problems, without having to run the millions of patterns need by random logic. What the test engineer is looking for is the composite response. With the above responses all occurring simultaneously, it is very difficult or impossible to separate them. The test engineer has the following variables at his disposal in verifying a design:

Cycle time

Temperature

Voltage

Pattern instruction set

Pattern repetition, duration, or "dwell"

Pattern weighting

Because SOI behavior is induced by specific use conditions, the test engineer has the ability to induce and repeat desired conditions, as convoluted as they may be, using the "knobs" shown above.

The use of Array Built-in Self Test (ABIST) has gained popularity as a means of streamlining and automating embedded array test. In bulk CMOS, a hard coded simple pattern could guarantee diagnosis of all but the most complex cache fail modes. The temporal variability in SOI may allow a handful of fail modes to elude ABIST detection using the conventional limited bulk CMOS pattern set. The solution is to enrich the test pattern set to encompass many machine cycles, in order to capture the full body voltage swings possible in their worst combinations. Given that this extended pattern set demands a more elaborate ABIST pattern generation engine, vulnerable array precondition scenarios must be carefully anticipated and included in the pattern set used.

Finally, long floating body time constants (which allow the bodies of MOSFETs in the array to reach equilibrium potential) can result in much longer SRAM test times, potentially 2 to 3 orders of magnitude compared to array test time in bulk devices. This time increase in most cases, is still small compared to the time required by the random logic testing on microprocessors.

6.9 Summary

The SRAM response to the SOI technology needs to be carefully examined, as each of SOI's idiosyncrasies appear in the SRAM's many structures. History effects appear in the word lines as variable latency, in the evaluate delay path as timing variation, in the cell as reduced noise immunity, and in the sense amplifier as bias. Reading a cell requires providing read delay which encompasses the worst case delay contributions of these SOI effects. Writing a cell requires anticipating the maximum peak currents caused by pass-gate preconditioning. These mechanisms must be anticipated in the design, as they can not be compensated for during operation. A multiplicity of responses occur simultaneously, and can not be separated by test. Luckily, SOI array structures lend themselves very well to a disciplined test approach which can uncover these problems. Performing these tests using built-in circuitry is more complex than in its bulk CMOS predecessor.

Not all the news is bad for SRAMs. Soft error upset immunity in SRAM, which has been steadily eroding from generation to generation in bulk CMOS, improves in SOI in a given lithography generation, as a result of a number of balancing mechanisms. While SOI soft error immunity erodes with lower voltage, it does so at a much slower rate than bulk. SOI has been said to love lower voltages: this observation is most readily apparent in the behavior of on-chip SRAM cache.

References

[6.1]   G. Shahidi, et al., "Partially-depleted SOI technology for digital logic," *Proceedings of 1999 IEEE International Solid State Circuits Conference*, February 1999, pp. 426-427.

[6.2]   D. H. Allen, et al., "A 0.2μm 1.8V SOI 550MHz 64b PowerPC Microprocessor with Copper Interconnects," *Proceedings of 1999 IEEE International Solid State Circuits Conference*, February 1999, pp. 438-439.

[6.3]   C.F. Edwards, W. Redman-White, "The effect of body contact series resistance on SOI CMOS amplifier stages," *IEEE Transactions on Electron Devices*, Volume: 44 12, Dec. 1997, Page(s): 2290 -2294.

[6.4]   P.E. Dod, et al., "Impact of Technology Trends on SEU in CMOS SRAMS," *IEEE Transactions on Nuclear Science*, Volume: 43 6 1, December 1996, pp. 2797-2804.

[6.5]   C.Lage, D.Burnett, T.McNelly, K.Baker, A.Bormann, D.Dreier, V.Soorholtz, "Soft Error Rate and Stored Charge Requirements in Advanced High-Density SRAMs," *IEEE Technical Digest of the IEDM*, 1993, pp. 821-824.

[6.6]   K. Bernstein, et al., *"High Speed CMOS Design Styles,"* Kluwer Academic Publishers, 1998. ISBN 0-7923-8220-X

[6.7]   M. Shoji, *"Theory of CMOS Digital Circuits and Circuit Failures"*, Princeton University Press, ISBN 0-0691-08763-6, 1992.

[6.8]   E. Takeda, et al., "A Cross-section of Alpha-Particle-Induced Phenomena in VLSI, *"Proceedings, International Electron Devices Society*, Volume: 36 11 2, November 1989, pp. 2567-2575.

·

CHAPTER 7   *Specialized Function Circuits in SOI*

7.1 Introduction

Aside from the implementation of combinatorial logic and cache memory in the SOI technology, as described in chapters 4, 5, and 6, other selected functions require additional attention to insure they will function as they did in bulk CMOS. Those circuits include SOI timing elements, latches, and input/output driver/receivers, and are described in sections 7.2, 7.3, and 7.4. Other specific circuits deserve attention, if only by virtue of the means by which they cope with the absence of substrate. Electrostatic discharge protection elements, described in Section 7.5, is one such functional unit. This book will not be discussing analog design techniques beyond the circuits mentioned above as it is beyond the scope of this text.

7.2 Timing Elements

Design of elements which synchronize events on chip must clearly accommodate SOI's history effect. In addition, microprocessors which use CMOS elements in an analog fashion must also be able to accommodate the other device phenomena outlined in Chapter 3 which are associated with DC currents. Those phenomena include self-heating and threshold voltage variability. Self heating was discussed in Section 3.6.1"Self Heating" on page 55.

Robust Phase-locked loop (PLL) design in CMOS is essential for high speed microprocessor operation. Analog PLL design in bulk as well as SOI CMOS depend on a few key structures, all of which are impacted by the move to SOI [7.2]. Those responses are summarized below. The reader is directed to citings in the literature for a review of PLL design fundamental concepts [7.1].

- Diodes
  Diodes are used in analog PLLs to establish band gap reference voltages. These diodes typically are formed by an n- diffusion placed in a p+ substrate. In SOI, the junction area is substantially reduced because the area component is lost to BOX. As a result, the perimeter diode must be increased, driving larger area dimensions for bandgap references in SOI.

- Capacitors
  Capacitors are commonly used in the charge pump section of the PLL to integrate phase differences, and convert them into voltage biases. The capacitors often were formed by n+ diffusions in n- wells, overlapped with n+ poly gates. In SOI the thin active silicon layer creates higher body sheet resistance, increasing the access RC time constant of the capacitor. The preferred practice, therefore, is to limit the aspect ratio of diffusions used to form these capacitors in SOI, to keep them narrow and quickly accessible.

- Level Shifting
  Shifting voltage levels in PLLs is conventionally accomplished by using the threshold voltage drop of an NFET or PFET. In bulk CMOS, the designer was stuck with a relatively fixed threshold voltage, perhaps varying it only via channel length. SOI offers an opportunity to reduce the threshold voltage by controlling the tuning device's body potential. In Figure 7.1, input V_{IN} to the uppermost NFET initiates a level shift, but the magnitude of that shift may be modulated by the voltage V_B coupled to that device's body.

- Analog MOSFET operation
  The use of the MOSFET as an analog element necessitates device stability. In SOI, this means making the PD-SOI device look as much like the bulk device as possible, in this case by tying the body to a fixed potential. Achieving good differential pair device matching is critical in the current mirrors comprising the PLL's comparator, as shown in Figure 7.2. To minimize input offset error, it is essential that devices N1 and N2 be closely matched. A floating body would allow body

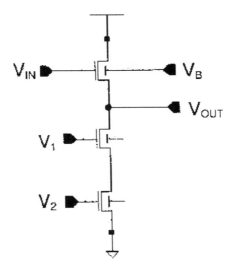

FIGURE 7.1 PLL Level Shifter. Magnitude of shift is modulated by input V_B.

charge accumulation in the devices to vary. Depending on the body contact technology, tying all PLL bodies to a potential also carries an area impact.

In addition, the body contacts are far from ideal. Current technology allows for connection on one end of the device, causing the contact resistance and the lateral body resistance to assert an RC delay which limits the effectiveness of the contact at high frequencies. Above a given frequency, the device I-V relationship changes because the body is effectively isolated from the rail it is tied to due to this impedance. The circuit output is then subject to increased jitter.

- Supply Noise
  In both bulk and SOI CMOS, great efforts are made to provide a clean quiet supply rail to the PLL circuitry, often by actually wiring in a separate supply. Nonetheless, logic synchronicity immediately succeeding a clock edge will induce noise in the PLL. The simultaneous switch of the latches at the clock boundary is somewhat reduced in SOI, as less capacitance is switched. This is countered however, by the loss of native N-well capacitance due to the BOX. The net effect is that SOI supply noise presented to the PLL is on the same order of magnitude as its bulk predecessor.

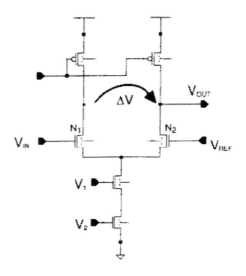

FIGURE 7.2 PLL Comparator, and matching of devices N1 and N2.

- History Effect in Clock Generator's common MOSFETs
  In microprocessors with high utilization, clocks typically are running continuously, and the bodies of these devices converge upon steady state switching potentials and remain there. History, therefore is not an issue in these applications, in which the clock has a constant activity factor. In selected desktop and in many portable computing applications, however, power-saving modes are implemented by gating clocks which serve a given logic macro. In these applications, the local clock regenerator as well as the logic structure sit idle for variable intervals, easily long enough to substantially alter body biases and clock delay when the circuit "awakens." Unless circuit structures are added to reduce this clock delivery variability, the timing of the part must account for this additional source of clock skew.

A phase locked loop (PLL) circuit must respond to asynchronous phase changes. The input stimulus could vary between one nanosecond or ten seconds. The phase detector should produce an identical response to all inputs, no matter when the input varies. With the variability of the device strength associated with the floating body, the switch current sources are the most sensitive to the devices history. The currents within these source can vary by more than 2X. To alleviate this problem, body contacts placed on the current sources that are switched are required.

7.3 Latch Response in SOI

The design of the registers holding machine state condition across cycle boundaries is easily one of the most critical components in logic products. Like the SRAM, the latch comprising the register circuit contains multiple elements, all known to be especially sensitive to SOI's nuances. Figure 7.3 shows the components of a standard L1/L2 master-slave latch circuit. Figure 7.4 shows the same circuit, highlighting how

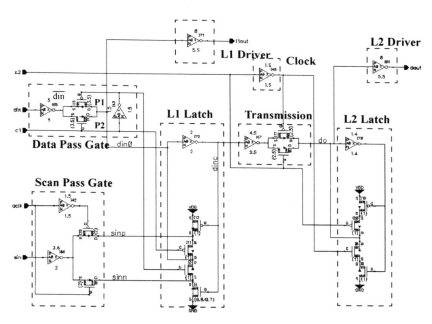

FIGURE 7.3 Standard L1/L2 Master-Slave data latch, and its major components

SOI affects the specific components. Those impacts are described in detail below.

7.3.1 Latch Stability and Noise.

The latch's central drain capacitances are reduced by the introduction of BOX, and diminish the noise immunity of the circuit. With less capacitance on nodes "din" and "sin" in Figure 7.3, conventional noise sources such as lateral coupling can disturb the contents of the latch; feedback strength delivered by the latch return inverters needs to be re-evaluated in SOI.

Specialized Function Circuits in SOI **153**

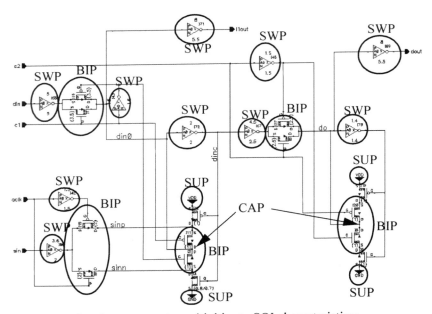

FIGURE 7.4 Latch component sensitivities to SOI characteristics:
SWP - Switchpoint variations
BIP - Bipolar parasitic current
CAP - Noise sensitivity due to junction capacitance reduction
SUP - Supply Rail Noise

In a similar manner to the SRAM, the high latch count on chip provided natural decoupling capacitance, which is eliminated in SOI. The distributed nature of the latches reduces decoupling capacitance over the entire chip. Supply rail-induced noise needs to be closely examined.

Finally, noise problems can be exacerbated by body charge accumulation in the pass gates. Even without the bipolar problem described in Section 7.3.2, higher pass-gate body potential results in lower threshold voltages, making them more susceptible to noise on clock lines.

7.3.2 Bipolar Charge Loss

Since most latches use pass-gates as the qualifying structure for the data into a latch, the same bipolar current which may disrupt a pass-gate can destroy the data held within a latch. Referring to Figure 7.3, if **din** is left high for many cycles and **din0** remains high as well, the body of devices P1 and P2 can charge to approximately V_{DD}. If signal **din** then goes high, **d̄in** goes low and charge is lost from node **din0** via bipolar current, even though select line **c1** remained high (inactive). Hardware data has shown that a latch with two write ports could be disrupted in normal operating space. causing it to lose it's data.

The design techniques for robust latch design are simple:

Provide more feedback strength in the latch to prevent the bipolar current from disrupting the data, or

Reduce the total pass-gate width to decrease the highest potential bipolar current for a given feedback strength.

Another simple fix for a latch with multiple data inputs is to guarantee that the signals on the passgate's input side do not arrive simultaneously, enabling the feedback within the latch to overcome the bipolar currents of each passgate serially.

7.3.3 Clock Skew due to History Effect

Aside from the latch itself, the variability of the clocks due to the history effect add a requirement that the hold time of the latch be extended to cover the effects of best case history and history variance from the clock generators. The history effect of the clocks can be considered worst when one portion of a clocking system remains running and another portion is gated to save power. When the gated portion is released, it will have a first switch characteristic while the main clock will be running with a steady state delay. This variability in clocks must be accounted for when timing the race conditions. In a $0.22\mu m$ technology, the variability of the clocks adds an additional variation of 25pS

Another potential cause for set-up and hold violations in the latch is not only in clock arrival, but also in how quickly the latch responds to the clock. History of the inverters comprising the latch can vary the threshold voltages and the resulting switchpoint of the inverters. Timing rules for latch components must assume the appropriate body voltages and thresholds for both set-up and hold times.

7.4 Input / Output Circuitry

Although the SOI mechanisms affecting driver and receiver circuits are the same as those impacting other microprocessor components, the operation of these circuits requires additional considerations. Analog timing elements, heavy output loading, and off chip synchronization requirements all exacerbate the SOI response.

Input/output circuits may be thought of as falling into four particular classes:

1. Clock Drivers
2. Data Drivers
3. Clock Receivers
4. Data Receivers
5. Voltage Reference

We discuss each of these in more detail below.

7.4.1 Clock Drivers

Simply put, the output of clock drivers are timed synchronized signals.

Among I/O devices, clock drivers are for the most part the least affected by SOI, largely because they run continuously, and so achieve and retain steady-state body potential. There will be little to no jitter or variation in output delay, due to the lack of history effects changing the clock driver output characteristics. There is little motivation to place body contacts on devices associated with clock driver circuits.

The one notable exception is when *clock gating* is used to quiesce functional units for power reduction. In this case, stability in clock phase and edge rate will not be achieved for many cycles, as the clock period converges back to steady state. This may range from 50 to 200 clock cycles depending upon the frequency of the clock and the accuracy require at the steady state value.

In both the steady state clock and the gated clock scenarios, use history affects both the body charge and junction temperature of the devices driving clock output. At steady state, the body potential delivers a fixed device performance, providing a fixed clock latency which also includes any temperature-related transconductance degradation. The temperature rise associated with consistent output should also remain stable.

As we will see below, the relative lack of history in the clock driver differs greatly from the history of the data driver.

7.4.2 Data Drivers

Data drivers differ considerably from clock drivers, Often drivers will output chip or array data in "bursts," outputting a "page" of data at a time. The given output may then sit idle for hundreds of nanoseconds. This usage pattern causes sharp rises in the device's temperature, as well as SOI history effect in the output device. These effects cumulatively result in first switch/second switch phase shift with respect to an accompanying delay line.

The present design practice is to provide output data along with a clocked delay line which will signal the receiving chip when data should be considered valid. Aside from normal clock skew and noise issues which affect the delay line, SOI history can assert 50 ps or more of shift in the arrival of data with respect to the delay line, which does not exit in the bulk case.

Body contacts are routinely used in SOI data driver final stages in order to minimize the variability in output delay or more specifically to avoid the variation on the output impedance. Large variation in the impedance will result in more noise on the data bus and require additional settling time for the signal. The body contact adds load to the output driver. The additional load comprises two main contributors:

More fingers of shorter width, resulting in additional parasitic capacitance

The body contact structure

This extra output load creates additional latency. This latency has been estimated to be increased by up to 50% of that of the previous design. The accompanying delay line is typically not body contacted. To achieve additional stability, often the pre-drive section is also body contacted to provide input to the final drive with as little variation as possible.

As we've noted previously, the present technology for body contacts is not ideal. Residual history effects exist in body contacted devices, and can generate residual jitter or variation in data rates. This jitter is minimized by reducing wide output drive devices into multiple short fingers which are well contacted.

> **Rule of thumb:** To minimize data jitter, limit the width of each driver device finger to not more that 4.5 microns in a 0.22 micron technology.

Given the issues surrounding data drivers in SOI, output data rates on SOI data drivers are approximately equal to that in bulk. Most of the data rate performance improvement is associated with the receiver, as we note below.

7.4.3 Clock Receivers

Clock signals, after passing the input through ESD protection devices and conditioning buffers, propagate into the on-chip clock distribution and local regeneration tree. In recent high speed designs, clocks are received differentially, and so are less sensitive to use history. If the clocks are never stopped at the system level, then the history dependence is also reduced. With wide differential input, clocks remain vulnerable to common mode noise, however, which remains unchanged from the considerations involved in bulk CMOS practice.

Current practice precludes the use of body contacts in clock receivers since the clock is free running and only level shifted to the appropriate voltage.

7.4.4 Data Receivers

Data receiver design is usually concerned with the selection of switch points and unity gain points, providing the circuit's noise immunity. SOI data receivers will exhibit more variability in its transfer characteristic due to history effect.

Nonetheless, because the receiver participates in each of the chip's cycle limiting paths, body contacts have not been used in the receiver. Experience has shown that the additional delay from worst case history is still less than the receiver delay had body contacts been included.

7.4.5 Voltage Reference

A voltage reference is used in multi-voltage I/O drivers to set an intermediate voltage on the pullup devices to limit the amount of stress on the PFETs. References are a DC analog device. Even though they do not switch, a floating body allows the threshold voltage of the device to reduces. Without controlling the body, the current through the device is increased and results in a large DC current on chip. To eliminate this, the devices are reduced in width making it more susceptible to variations in width control by the process. To get around variable voltage references and large DC currents, the use of body contacts is considered prudent.

7.5 Electro-Static Discharge (ESD) Protection

Don't judge a book by its cover!

The presence of a buried oxide abutting source and drain diffusion implants has been inappropriately judged as a liability in providing acceptable ESD protection. As disclosed in recent literature, SOI can be shown not only to not reduce ESD protection levels, but to offer a number of advantages over its bulk predecessor [7.3].

1. ESD protect diodes, in most common configurations, present additional junction capacitance to the driver. With the area component of the ESD diode greatly reduced, the resulting load presented is smaller.

2. The combination of shallow-trench isolation and buried oxide provides an enhanced means of reducing sensitivity to layout indiscretions. With the ESD structure surrounded by dielectric, normal hazards such as Human Body Model (HBM) discharging, latch-up, parasitic current paths, and charge injection are avoided.

3. ESD layouts may be made more compact in SOI. The absence of guard rings, and of minimum spacings between P+/N+, n+ to guard ring, resistor to guard ring, resistor to resistor, ESD to I/O, and resistor-to-I/O insure denser layouts and easier technology migration.

4. The presence of the BOX specifically allows for innovative body-tied ESD protection schematics.

5. Finally, the BOX eliminates the charged-device-model (CDM) body-to-gate failure mechanism in devices with floating bodies.

The *lubistor* transistor diode configuration has been used successfully to suppress ESD-induced failures in recent high speed SOI microprocessors. This Lateral Unipolar Bipolar transistor structure, shown in Figure 7.5, uses a gate electrode to replace

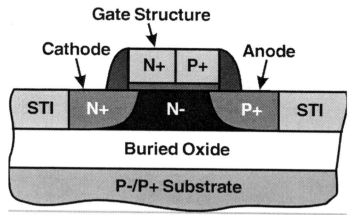

FIGURE 7.5 Lubistor protection structure as practiced in SOI designs.

an STI region. It has been shown that ESD protection voltages increase linearly with diode perimeter. The structure improves protection with scaling, due to the reduced diode series resistance at shorter physical poly line width. An SEM photomicrograph of the fabricated structure highlighting the structure is shown in Figure 7.6. Failure

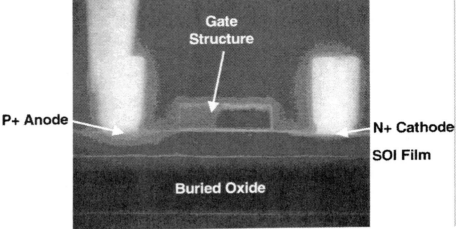

FIGURE 7.6 SEM Photomicrograph of Lubistor ESD Structure. (Photo compliments of IBM Microelectronics)

testing data has shown SOI ESD structures to provide higher protection levels than bulk CMOS structures in some scenarios.

Structures such as the dynamic threshold body-coupled and gate-coupled circuit shown in Figure 7.7 are unique to SOI, and can enhance protection.

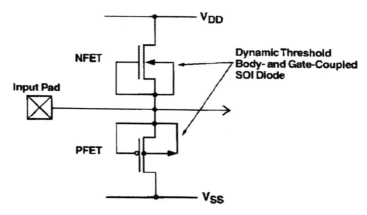

FIGURE 7.7 Dynamic threshold body-coupled and-gate-coupled SOI ESD circuit

7.6 Summary

The implementation of logic in a high speed CMOS technology as complex as PD-SOI requires not only robust circuit design practices, but substantial behind-the-scenes supporting infrastructure. This overhead, while it executes little logic function, assures that the microprocessor's external interactions are compatible with the anticipated timing and technology environment.

Circuitry which keeps the microprocessor synchronous with the overall system makes substantial use of the MOSFET's analog properties. Because SOI's floating body asserts temporal variability, adaptation of the SOI device to analog applications requires the designer to mimic the bulk device as much as possible. Exploiting body contacts, minimizing body resistance, and altering offset potentials are a few of the new design priorities of timing circuits in SOI.

Latching elements, which trap the intermediate results of the state machine, display a profound response to SOI. The change is mainly due to the "good" capacitance reduction and the potential for low threshold voltages. Increased noise sensitivity is the most significant concern, and feedback in the latch must be increased to compensate for the loss.

The burst nature of output driver usage makes I/O circuitry especially vulnerable to history effects and self-heating. The use of body contacts, which has been known to add delay, is avoided in clock I/O because the latency is stable, and is often omitted in data I/O because the delay may exceed the history variability.

Finally, electrostatic discharge protection structures in SOI are quite different from their bulk predecessors. To compensate for the loss of junction area now occupied by BOX, innovative schemes have been developed to enhance the current sunk per micron of perimeter diode length.

So, since we now have developed an understanding of the SOI logic and support circuit electrical issues, all that remains is to effectively deal with resulting global chip design issues arising during the product's physical image[1] definition. We address these concerns in the next chapter.

---

1. *Image* refers to the conventions adopted in the product for power and clock distribution, standard cell aspect ratio, circuit placement, dataflow bit width, array proportions, etc.

REFERENCES

[7.1]   Floyd M. Gardner, *Phaselock Techniques,* John Wiley & Sons, Inc., New York, 1979.

[7.2]   J. Eckhardt, et al., "A SOI-Specific PLL for 1 GHz Microprocessors in 0.25 μm 1.8V CMOS," *Proceedings of 1999 IEEE International Solid State Circuits Conference,* February 1999, pp. 436-437.

[7.3]   S. H. Voldman, "The Impact of Technology Evolution and Scaling on Electrostatic Discharge Protection in High-Pin Count High-Performance Microprocessors," *Proceedings of 1999 IEEE International Solid State Circuits Conference,* February 1999, pp. 366-367.

CHAPTER 8 *Global Chip Design Considerations*

8.1 Introduction

The insight required in the crafting of a high speed processor is nowhere more important than in the consideration of design parameters which affect the product at a global level. Physical chip design must accommodate the range of temperatures the product may see; the noise a generic circuit might be exposed to; the power the design might consume; and the variation in the performance of individual design components caused by process, layout, prior use, or wearout. The total system performance advantage realized by a design in a given technology is strongly dependent on the design quality of the power distribution, clock distribution, signal wire coupling, and technology selection, all global design parameters addressed during chip physical design[1].

Given the foundation built in the previous chapters, we now examine chip PD issues using this new insight.

---

1. a.k.a. *Chip PD*

8.2 Temperature Effects

Temperature affects the SOI MOSFET in a number of ways. To a first order, the SOI MOSFET response to temperature is similar to that of a bulk MOSFET. Differences arise from the SOI MOSFET's BOX-isolated body, and its dependence on the behavior of its source and drain junction diodes. Depending on the temperature window, the results can be surprising. Normal operating temperature, stress temperatures, and low-temperature SOI design responses are described below.

8.2.1 Design Considerations, Normal Operation

As described previously, the temperature variation caused by AC device operation in SOI asserts less than a 2% delay variation due to self-heating. Devices passing linear or saturation currents for extended periods of time, however, can develop large temperature differentials, causing dramatic swings in transconductance. For that reason, a few SOI principles significant to circuit design are listed:

1. **Temperature Isolation**: Silicon is a good heat conductor. For that reason, if an active FET is expected to have large currents for extended periods, it should not share its diffusion area or body with another device in a critically-timed path which would also lose transconductance, as shown in Figure 8.1. This plot illus-

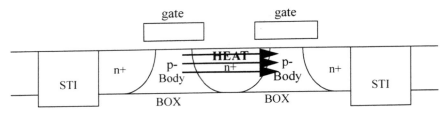

Active MOSFET   Victim MOSFET

FIGURE 8.1 Temperature influences on neighboring victim devices

trates the cross-section of a pair of MOSFETs. The heat from the active MOSFET is transferred to the victim MOSFET. This will reduce the mobility of the victim MOSFET and degrade its current carrying capabilities.

2. **Matching**: On the other hand, if a particular design requires the close matching of two transistors (i.e. current mirrors, sense amplifiers), placing them in the same orientation and in close proximity are beneficial. In SOI, placing both transistors

in the same silicon island can reduce variation. The designer must assess the relative gain in tolerance control by matching, vs. the loss in tolerance control from heating.

3. **Heat Sink:** In places where the currents are high, but the loss of transconductance is undesirable, silicon and/or thermally conductive contacts can be used as a heat sink to conduct away some of the heat. This would be accomplished by expanding the diffusion areas and adding more contacts to metal above the device. Since the FET is surrounded by silicon dioxide, common heatsink metal materials still serve as the best external heat sinks.

4. **Insulation:** The circuit designer should remember that not only the BOX below the transistor, but the STI to each side as well as the inter-layer dielectrics insulating the wiring levels are also made of silicon dioxide. Silicon dioxide is a poor thermal conductor, and the transistor has it to the left and right (STI), below (BOX), as well as above (wire interlayer-dielectrics). In essence, the transistor sits in a perfect "cooler," making it harder to dissipate heat. This impediment also asserts a shorter-term use history effect by more quickly reducing majority carrier mobility with temperature.

5. **High Activity Factor:** If a given circuit with a relatively low switch factor is placed amidst a larger functional block with high activity or linear mode circuits, the SiO_2 "integrates" the heat generated, raising the ambient temperature for that device, even though it may not switch. Larger area effects are strongly dependent on the nature of the instruction stream and the proximity of active devices, making thermal analysis quite a task.

8.2.2 Stress Considerations

Designs going into applications with high reliability requirements are often stressed at elevated voltages and temperatures to accelerate the exposure and fallout of manufacturing defects which would have otherwise occurred in the field. For these applications, one way of guaranteeing the exposure of each node on the chip to full stress conditions is to design the product to remain functional at these "Burn-In" conditions. Burn-In and screening operating conditions may be at temperatures of 120°C or higher, using supply voltages exceeding 1.5X the normal operating voltage. They are usually operated at very low clock frequencies (1-5 MHz) to minimize equipment costs.

1. Electrical Response - Functionality during stress

At stress conditions, MOSFET bodies will experience much greater swings in potential, creating greater delay variability. This is exacerbated by the increased edge slew rates caused by higher V_{DD}, although countered by reduced transconductance at high temperature. First-switch/second switch offsets are likely to diminish quicker at a given pulse width, as the junction diodes are leakier and will act to stabilize body voltage sooner.

The higher supply voltage will result in higher body potentials which in turn lead to larger passgate bipolar currents. If functionality is to be maintained, then the larger bipolar current should be accounted for in design margins.

The increased temperature and increased supply voltage that increases the body potential and increases the leakage of the drain to body diode will result in a large increase in the parasitic bipolar leakage current. This leakage will increase the leakage current that is present under the stress conditions. This will be covered further later in this chapter.

2. Physical Response - Thermal Runaway

The superior thermal conductivity of silicon essentially prevented bulk-substrate CMOS designs from developing excessive thermal feedback. SOI, on the other hand, demonstrates a lower finite power limitation which may not be exceeded. Figure 8.2 portrays the operating space for a recent PD-SOI test chip, displaying thermal instability in high-current operating corners [8.1]. As device temperature increases, thresh-

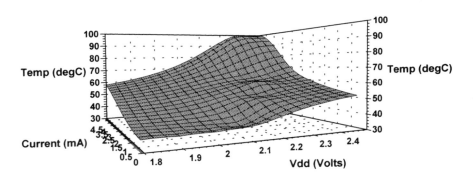

FIGURE 8.2 SOI Thermal Runaway at stress conditions [8.1]

old voltage is reduced, which increases leakage currents and enables a positive feedback mechanism. Thermal runaway is said to occur when the temperature of a product module no longer is bounded to a given upper limit, but steadily increases with time. This increase in temperature can occur rapidly (i.e. within seconds). Thermal runaway is destructive, and is a concern both at stress and in insufficiently cooled product applications.

> **Rule of thumb:** Under stress conditions, the AC component of power is negligible and the total power can be calculated using the leakage components on chip.

To alleviate the possibility of thermal runaway during stress one has two choices. First, the applied stress voltage could be reduced and prevent such large currents from being generated. This will result in longer stress times needed to obtain the equivalent acceleration of the stress. Second, if the current is large, then good thermal control must be provided. This may be possible by individual thermal control of each module under stress. This option requires additional cost on the stress testing equipment, but alleviate the possibility of long stress times.

8.2.3 Low Temperature Operation

Processors realized in PD-SOI and operated at low temperature may be thought to be designed to two distinct set of considerations. They are outlined below.

1. Fundamental CMOS Low Temperature Behavior

CMOS exhibits characteristics at lower temperatures (say $0^{\circ}C$ and below), which diminish as the product is moved to normal operating conditions of $70\text{-}100^{\circ}C$. Most of the noteworthy behavior at lower temperature is associated with the relative increase of both device transconductance and device threshold voltage. Figure 8.3 and Figure 8.4 provides the response of a N-MOSFET's $I_{D(SAT)}$ and threshold voltage to temperature in a $0.20\mu m$ technology. Depending on process implementation, one should not necessarily assume that NFET and PFET devices will track in their improvements. The CMOS technology's composite behavior needs to be anticipated

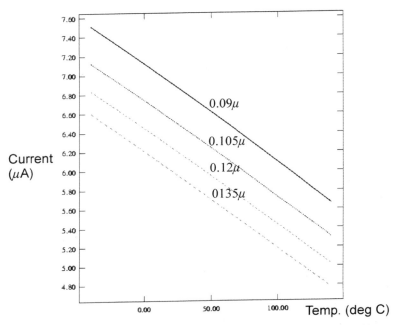

FIGURE 8.3 N-MOSFET I_{DSAT} dependence on temperature for 4 different channel lengths in a conventional 0.20μ technology.

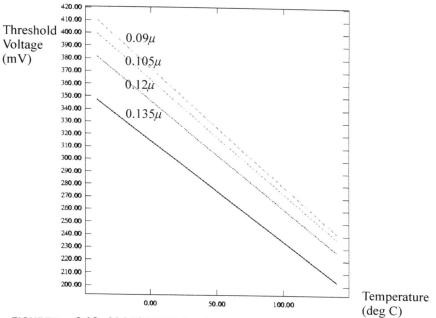

FIGURE 8.4 0.18μ N-MOSFET threshold voltage dependence on temperature, for 4 different channel lengths in a conventional 0.20μ technology.

in the design. This design issue is *independent of the use of PD-SOI*. Design considerations which must anticipate the use temperature range include:

Capacitive Gain / Stage ("Alpha Ratio")

Device strength Ratio ("Beta ratio")

Selection of Switch Point / Unity Gain Point

Interconnect Performance (Thermal Coefficients for Resistivity)

Logic Noise Immunity (Asynchronous Noise)

Delay Noise Immunity (Synchronous Noise)

Latch Setup / Hold Allowances

SRAM Transfer Ratio / Cell Sizing

SRAM Soft Error Upset changes

Analog Element Design (i.e. PLL)

Driver/Receiver Impedance Matching

Special Function circuits (i.e. Electro-Static Discharge Protection)

Clock and Power Distribution

Power Consumption

Accurate Transistor Models

Device and Interconnect Reliability

System Thermal Cycling Limitations

Condensation

These fundamental design considerations must be treated in bulk CMOS technology as well as SOI; while mentioned for completeness, these issues are beyond the scope of this text. The reader is directed to the referenced literature for a thorough assessment of CMOS at reduced temperatures [8.2].

2. Low Temperature SOI Device Behavior

SOI-specific low temperature issues are associated with changes in the body potential of the device, and the resulting history effect and threshold voltage movement. At reduced temperature, both the source-body and drain-body junction diode leakage reduces, resulting in a DC equilibrium potential of the body of the NFET that is higher than warmer temperatures. With a nearly identical amount of capacitance on the junctions that touch the body, this increased potential narrows the performance delta observed in first and second switch and reduces the history effect.

Note to the reader: As mentioned earlier in this text, the first and second switch ratios are a specific to a given device design point. It is possible to have the second switch slower than the first switch. In this case, the DC equilibrium potential of the body is larger than the amount of AC coupling to the body by the capacitance connected to the body. In such a case, as the temperature decreases, the DC equilibrium potential continues to increase and increases the history effect. In this text, we have assumed one design point of second switch being faster than first switch for most discussions.

Figure 8.5 shows the delay down a long scan chain[2] that has been opened in flush

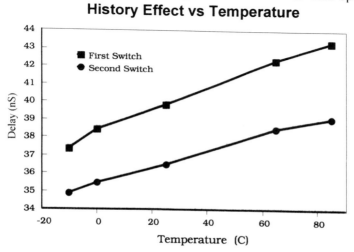

History Effect vs Temperature

FIGURE 8.5 Scan chain flush delay as a function of temperature. First switch and second switch are shown.

mode. Here a rising input was presented at the beginning of the chain and each latch was held open and the delay was measured from input to output for the first and second switch. As is apparent in the graph, the delay of the flush chain is decreased as the temperature is decreased for both first and second switch, since the mobility of the carriers in improved. The amount of history effect (i.e. the difference between the two curves) has decreased. If the temperature is reduced farther, the two lines would inter-

---

2. *Chain* refers to a sequence of circuits, in this case inverters, attached serially: the output of one element serves as the input to the next in the chain. For inverting elements, an **odd** number of elements chained into a *closed loop* will produce *ring oscillation*, which can be measured at any point in the ring using attached counters.

sect and there would be no history effect. To reiterate, if the second switch had been faster to start with, then a decrease in temperature would have increased the amount of history effect as the slope of the second switch is shallower than the slope of the first switch.

Recent literature provides additional insight into the related effects of temperature and activity factor on history magnitude [8.3]. Above, we learned that first switch and second switch performance can be a strong function of temperature. We observe in Figure 8.5 that the change in rising and falling output performance due to history and temperature should not be assumed to be equivalent. We see that in going from single first switch events to steady state events which have achieved steady state body potentials, falling output transitions may decrease in variation, while rising output transitions can increase in variation.

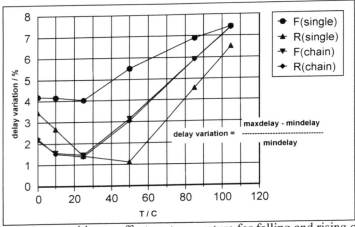

FIGURE 8.6 Inverter history effect vs. temperature for falling and rising output edges, for a single inverter and a 51-inverter chain [8.3]

Further, the length of the inverter chain establishing the steady state also can impact the performance variation. In Figure 8.7, the delay variation dependence on chain length is explored for the falling output edge, in 0.22μ technology operated at 1.8 V. At all temperatures, delay variation increases with decreasing chain length, due to the reduced time allowed for the devices to decline in body potential during and between events. The temperature influence is seen in the magnitude of the history effect on the falling output at short chain lengths, as at higher temperatures the bodies lose their excess potential more quickly.

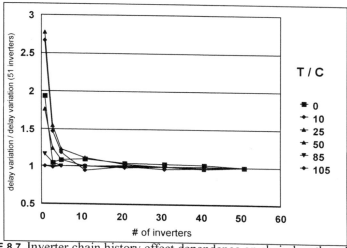

FIGURE 8.7 Inverter chain history effect dependence on chain length and temperature [8.3]

8.3 Noise Immunity

Noise in SOI chips is similar to bulk, with a few additional considerations. In this section, the definitions of different noise events will be reviewed. The deltas between SOI and bulk will be explored, but the presentation of noise on all types of circuits will not be presented. This work was presented for CMOS circuits in [8.4]. The types of noise that are apparent in all circuit designs are:

Cross Talk

Delay Noise

Logic Noise

Simultaneous Switching

The moment when a critical noise event happens is most of the issue with the immunity. Noise may occur at the clock edges when the latches fire. This type of noise is synchronous noise. When a noise event occurs without any apparent timing relation to other events, these are called asynchronous noise.

8.3.1 Crosstalk (Capacitive Coupling)

Any two metal wires that lie next to each other are electrically isolated by the dielectric that lies between them, but are electrically connected by the capacitance across this dielectric. The amount of current that is exchanged by the two wires is dependent upon $i=CdV/dt$, where C is the capacitance between the wires and dV/dt is the rate of change of the voltage between a wire and its neighbor. This injected current creates a voltage pulse upon one wire (the victim) when another wire (the aggressor) switches. This is the essence of crosstalk noise.

The amount of capacitance is a function of the width of the wire, the linear space between the lines, the thickness of the wire, the separation distance to any other conductor such as wiring levels above and below, the type of dielectric and the length of the common run shared by the two wires.

SOI is a technology in which circuits switch faster. Therefore, the edge rate of signals is faster and more current is injected across the coupling capacitor. Not only must one handle more injected charge on the victim wire, but the amount of capacitance on the victim wire is sure to be less than in bulk since the junction capacitance has been reduced. The amount of coupling capacitance is nearly identical since the interconnects do not change with respect to SOI on the upper layers of metal. Therefore, the ratio of coupling capacitance to the wire capacitance is higher resulting even a larger voltage pulse on the victim wire.

8.3.2 Delay Noise

When induced noise from the aggressor wire either speeds up a victim wire or slows it down, this is called delay noise. It is synchronous in nature since both the victim and aggressor net are switching at the same time. If the victim net is on an upward transition, and the aggressor net is also on an upward transition at the same time, then the dV/dt is very small and the driver of the victim net will have less effective capacitance to charge and this will speed up the driving circuits delay and reduce the RC delay of the victim. This may induce a race condition if the decreased delay causes the signal on the victim net to arrive too soon at its destination circuits. For example, consider a data path that is sped up due to coupling and arrives at a latch, before the latch is closed to prevent the current correct data that is within the latch to be maintained. This is a failed race condition where same direction switching can cause chip failures.

The counterpart to this example is when the victim net and the aggressor net are switching in the opposite directions. Here the voltage change across the coupling capacitance is now 2dV/dt and results in a larger effective capacitance for the driving circuit to overcome. This added delay could result in a signal being propagated too slow and failing to arrive at a latch in time to overwrite the latch with the correct data. Performance limited paths that consist of a large amount of RC delay must accommodate the added delay due to switching neighbors to accurately predict a chips final cycle time.

8.3.3 Logic Noise

Speed variations from delay noise is not the only issue with crosstalk, logic noise and functionality are of concern too. Logic noise is defined as an asynchronous noise pulse from a neighboring net onto victim net which is quiet which causes a circuit to fail. The forward propagation of this logic noise onward into a cone of logic is of utmost concern because chip failures will result if an incorrect value is captured within a latch.

An inverter will pass a large enough pulse through it incorrectly, but then return to the correct state due to its static nature. Figure 4.11, "Noise Schmoo for an inverter in SOI and bulk CMOS.," on page 80 showed what size input pulse was required to create a pulse at the output of the inverter. If this pulse occurs at the wrong time, then the incorrect data would be captured at a latch. This may not be a total failure, since with static circuits, if the frequency is slowed down, the correct value will eventually be latched.

Other circuit families such as dynamic circuits and passgate circuits are not able to recover from some logic noise fails. If a large noise pulse is presented to a dynamic gate that triggers the gate, it will not recover. With floating body effects, the reduction of threshold voltages directly results in a reduction of noise margin. It would now take a smaller noise pulse to cause a dynamic or passgate circuit to fail. Remember that bipolar currents were discussed in Chapter 5 with dynamic circuits. Now consider what happens if a noise pulse is presented at the gate of the device that is passing a bipolar current at the same time. A synchronous noise pulse will turn on the device creating a larger amount of charge pulled off of the dynamic node by both the drain-to-source current and the bipolar current. This is an SOI unique situation and reduces the noise margin even farther than the reduce threshold voltage alone has done.

It is possible for noise to add across several stages of logic. If the gain of several circuits is large enough, the output of one circuit will produce a noise pulse that is larger than the input. When the output noise pulse is then presented at the input of another circuit with a large gain, its output will be larger. Over several stages, this noise pulse will continue to rise until it creates a logical fail. This can occur naturally for dynamic adders, where the footer device is turned on and produces a bipolar current simultaneous with the noise event.

8.3.4 Simultaneous Switching

Simultaneous switching of neighboring circuits cause simultaneous need for current to be supplied by the power distribution grid. This can cause V_{DD} power rail to droop during the switching events. This is most prevalent with I/Os and the power supply associated with the I/Os. Usually the I/O's power rails are kept separate from the internal chip power distribution system for two reasons. First, the I/Os may have a different output voltage than the internal voltage of the rest of the chip. For example, microprocessors in current technologies have internal voltage of 1.8 V and the I/Os must interface with a system board that is at 3.3V. Secondly, the added noised that I/Os generate on the I/O's power supplies is kept off of the internal power distribution system if these rails are kept separate. This is good for the internal circuits since there power supply noise is now reduced. However, this is a disadvantage for the I/O's power supply since there distribution system is now smaller since it is kept separate from the rest of the chip and there is much less decoupling capacitance for it to draw upon. Another problem with I/O power supplies is that when simultaneous switching occurs, the demand upon the distribution system are quite large since each I/O need to drive the very large capacitance of the system. The local voltage of an I/O can droop as much as 25% when all of the I/Os are switching at one time. To reduced the amount of droop on the I/O power supply, one typically will limit the logic from switching all the I/Os at once, or will add internal decoupling capacitors within the chip or external decoupling capacitors to the module.

8.4 Power Consumption

SOI has been touted over the years as a great process for reduced power consumption due to the reduction of total capacitance on a chip. Equation 8.1 shows the relation between power (P), capacitance (C), voltage (V), frequency (f) and switching ratio (s). Power savings come mainly via two opportunities. First, the total capacitance has been diminished due to the reduction of the diffusion capacitance. Since the junction capacitance is not the only capacitance on the chip, this results in approximately a 10% power savings compared to bulk. The remaining components of total chip capacitance, (i.e. interconnect capacitance, gate capacitance, overlap capacitance) are still prevalent in SOI. Secondly, with the performance advantage possible with SOI, it is possible to operate the circuit at the same frequency, but at a lower voltage. Since the power-voltage relationship has a square law dependence, a small reduction in operating voltage generates a larger savings in power. It is feasible to operate a given product with a power savings of 70% at an identical frequency by reducing the voltage.

$$Power = sCV^2f \qquad (8.1)$$

Figure 8.8 shows this use of reduced voltage as a method of power savings. At an

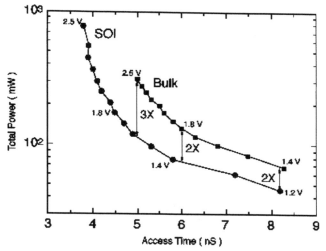

FIGURE 8.8 Power vs. access time for an SRAM at different voltages.

access time of 5nS, the power has been reduced from 300mW to 100mW, since the power supply could be reduced from 2.5V to 1.6V, resulting in a 67% power reduction. Under a similar vain, the access time has been reduced by 25% from 6nS to 4.5nS at 1.8V. Whichever use of SOI is most applicable for your application, one may trade-off power vs. performance.

SOI does have a downside to the power situations. With the floating bodies, the threshold can become very low, which of course is good for performance, but the crossover current has increased. In an inverter, the peak current has increased by approximately 15% when the input has transitioned through the switchpoint of the inverter. This crossover current is very short lived, but it does add to the total AC power consumption.

Also, with the floating bodies, the leakage of the off device has increased. In addition to the substantial FET leakage, the parasitic bipolar device is in its conducting state with the body floating. This results in an increase in bipolar current that is not existent in the bulk technologies. At higher V_{DD}, the bipolar device is turned on more and more. With a voltage near 2.5-3.0V, the bipolar's base is being controlled by the impact ionization current from the subthreshold leakage current. See Figure 3.20 on page 51. This extra charge in the base is turning the bipolar device on and the bipolar transistor starts to break down. Multiplied over millions of FETs, the current is easily observable. Figure 8.9 shows how the sleep power is a function of V_{DD}. Since this is

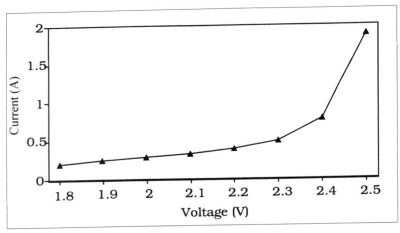

FIGURE 8.9 Sleep current vs. V_{DD}

a plot of sleep power, the power consumed is solely due to leakage currents. As the voltage is raised, the power starts to climb dramatically near 2.4V. Since sleep mode mainly consists of leakage current, here are two rules of thumb for determining the amount of leakage.

> **Rule of thumb:** To determine the amount of leakage in an array, use an average device width and multiply by the number of device. Finally, assume that only 50% of these devices may leak at one time.

> **Rule of thumb:** To determine the amount of leakage in the logic portion of a chip, use an average device width and multiply by the number of device. This time, assume that only 33% of these devices may leak at one time.

Finally, the total leakage is a weighted average of these two components. For small microprocessors with a modes amount of memory, the percent leaking is 40-45%. For chips with large integrated caches, such as L2 caches, this number may approach 47-49%.

> **Rule of thumb:** When calculating the leakage current, never forget to include the component due to ACLV effects. The short channel devices provide the majority of the current.

As technology scaling decreases the voltage farther and farther, the operating range will be well away from the 2.5V range. One issue that will continue to remain will be the high voltage screens put in place for finding defects and improving the reliability screening.

To further prove that the current is from leakage and not from poor control of the sleep current, Figure 8.10 shows a backside optical emission of the current flow within a microprocessor using a two probe technique contacting V_{DD} and GND only. The bright spots within the image are locations were the current is flowing. The image was collected with a two amp limit at about 2.6V. The I/Os were left floating and this section of current flow is expected. The other regions that are lit up are banks of inverters and muxes. The muxes contain passgates and inverters. Finally, the L2tag labeled on the bottom of the photo is an array of SRAM cells that are simply cross coupled inverters. Therefore, the current flow appears to be coming from single high stacked devices, namely inverters. The only way to get rid of this current at a given voltage is to have a process change that moves the onset of the bipolar device higher and away from the operating or high voltage screen voltage. The other option is to operate the device away from the higher voltage range. At a reduced voltage, the bipolar currents are small and the current consumption and hence the power savings is large.

In SOI, the FET is surrounded on all sides with an oxide which is a poor thermal conductor, this results in some localized heating or self heating which elevated the temperature of the local devices. As mentioned before, this may result in a possible

FIGURE 8.10 Optical image of the current flow with two probe contact.

unstable condition known as thermal runaway if the thermal cooling of the chip is not adequate. Figure 8.11 shows the current increase as a function of time at a given voltage. At each voltage plotted on the graph, the current was measured as soon as the voltage was applied, 15 seconds later, and 30 seconds later. The current was increasing over time as was the temperature of the chip (not shown) was increasing. At 2.1V, the thermal runaway was beginning to start, but even after 30 seconds, the current had not increased much. At 2.4V, the current had increased 40% in 30 seconds. Above 2.4 volts, the three amp compliance limit was quickly reached.

In this text, we have talked about the possibility of the substrate acting as a "backside" gate for the SOI devices. Several issues arise with this backside device. First, the thickness of the BOX leads to a high voltage needed to turn the device on. If the substrate is raised high enough, it will turn the parasitic backside FET on. Not only will it turn this one parasitic backside NFET on, but it will turn nearly all backside NFETs on at once, which will lead to an large increase in "leakage" current. Again, in a similar manner, the substrate can act as a gate for a backside PFET for the PFET

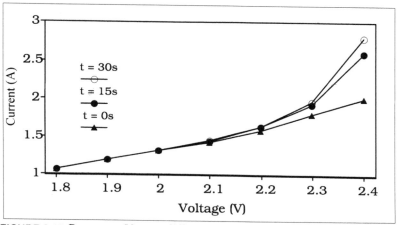

FIGURE 8.11 Power vs. V_{DD} at different time intervals showing thermal runaway

structure. In this case, the substrate would have to be driven to a large negative voltage to turn on the PFETs, but if the substrate is floating, this is possible. Just touching the substrate can change the potential enough to activate the backside devices. If the substrate is not connected, then the substrate will hold its charge for a very long time. To remove the backside issues, the substrate should be connected to GND with a connection through the BOX or with a direct connection to the backside. With a floating substrate, it was found that touching the backside of the chip would cause a passing chip to fail. Even after waiting several minutes, the chip would not work. Touching it again would change the amount of charge on the backside and change the current consumed by the microprocessor. Figure 8.12 shows the leakage current as a function of the voltage on the backside under more controlled conditions.

As the backside of the chip was raised to a positive voltage, the NFETs would start to turn on since the substrate was the "gate" and the BOX was the "gate oxide". As the voltage was set more negative, the current consumed by the microprocessor again increased as the PFETs were turned on by the backside gate. As mentioned before the functionality was impacted as the voltage on the backside was changed. Above 5 volts and below -9 volts, the microprocessor failed to function properly.

Work was done to create a contact between the GND on the active side of the chip and the substrate. This required a connection through the BOX as shown in Figure 8.13. This connection is allowed to be highly resistive, however, if one creates a contact ring around the edge of the whole chip the resistance is not too large and will maintain

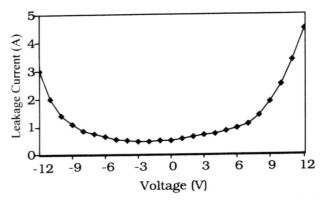

FIGURE 8.12 Leakage current as a function of voltage on the backside of the substrate with the substrate floating.

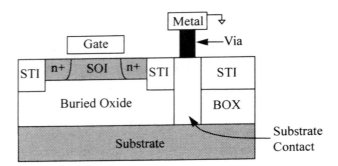

FIGURE 8.13 Substrate tied to GND with a backside contact.

the backside gate at GND. This backside connection is very similar to the guardrings that are created on bulk chips, only it controls the backside gate, not the body of every NFET.

8.5 Power Supply Issues Noise

As has been mentioned before, the insertion of the BOX has reduced the total decoupling capacitance due to the loss of the natural NWELL to substrate capacitance that exists in bulk technologies. On a small microprocessor, about 40 mm^2, this resulted in a loss of about 7.5nF of decoupling capacitance. This was thought to be about one-third of the total on chip decoupling capacitance. With this reduction in decoupling capacitance, the noise on the supply rails is increasing. In additions to the decreased decoupling capacitance, the chip is capable of running faster and has faster slew rates resulting in even more noise on the power supply.

Figure 8.14 shows the supply rail compression for three different amounts of capaci-

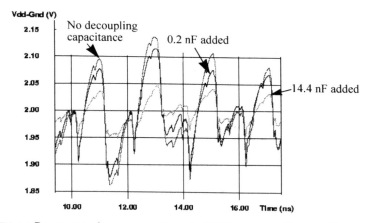

FIGURE 8.14 Power supply compression of a 2.0V supply with no additional decoupling capacitance, 0.2nF or 14.4nF of decoupling capacitance.

tance on a microprocessor. With no additional capacitance, the supply rail moves about 130mV above and below the nominal supply of 2.0V. When 0.2nF of capacitance is added close to this position on the chip, the noise is reduce to 120mV. Finally, when 14.4nF, which is a large amount was used to fill in additional white space on the chip, the total noise on the power supply was reduced to about 50mV.

Less noise is better, but under what conditions should one spend more time or more area to reduce the power supply noise. In terms of performance, let's consider both full cycle and half-cycle paths. A full cycle path would see additional performance gain when the voltage rises above 2.0V and speed up, but it would have reduced performance when the voltage drops below 2.0V. Since the period of peaks are determined by the clock frequency of the microprocessor. Each full cycle path speeds up and slows down. Averaged over the full period, the delay of a full cycle path is not changed from a power supply that has no noise.

A half cycle path is a different story. If the half cycle path occurs during the power supply peak, then the delay is improved. However, half cycle paths that occur while the power supply is compressed, the delay will be increased. A decrease of 100mV on the power supply during this half cycle can result in a decrease in performance by as much as 5%. Clearly a less noisy supply can help solidify the performance of a microprocessor as many half cycle paths exist on each chip.

8.6 System Performance

SOI has shown great performance advantages at the circuit level. At the chip level, the performance is reduced. Although device performance is improved, chip performance is comprised of gate delays and interconnect delays. SOI did not improve the interconnect delay by much. Minor improvement in delay for the interconnects is seen in the first level of metal, since the substrate is moved further away from the metal by the buried oxide, the total capacitance on the first layer of metal is reduced slightly. The second layer of metal is shielded from the substrate by the first level of metal and most custom chip designs have the large majority of the interconnect delay on the higher levels of metal. Therefore, one should not count on any improvement of the interconnect delay due to SOI.

In Chapter 4 and Chapter 5, the individual circuit delay improvement was shown to be 15-48%. For a simple example, let's assume that the circuit level improvement is 35%. If the interconnect delay is 20% of the overall timing budget and SOI improved the device delay by 35%, then the overall system performance improved by 28%. As interconnect delay is a smaller portion of the critical timing paths, then the gains from SOI approach the 35% device performance gains. Figure 8.15 shows the delay and frequency improvement for larger custom macros within different microprocessors. The range of frequency improvement was 24% to 43%. In this figure the impact of large wire load is apparent for CPU2 and CPU3. To take full advantage of the performance possibilities, the amount of delay due to interconnects should be reduced through judicious floorplanning or metal thickness and profiles.

Circuit	$\dfrac{t_{Bulk}-t_{SOI}}{t_{Bulk}}$	$\dfrac{f_{SOI}-f_{Bulk}}{f_{SOI}}$
CPU1 L2 Directory Access t	29.8%	43%
CPU1 CP	22.6%	28%
CPU2	27.1%	37.1%
CPU2 (high wire load)	20.9%	26.4%
CPU3	30.4%	43.6%
CPU3 (high wire load)	19.6%	24.3%

• Using same generation bulk and SOI technology

FIGURE 8.15 Performance gain for larger circuits. Both delay and frequency improvement are shown.

As with any technology, someone wants more performance. The usual methods are to decrease the channel lengths. In bulk, this gives performance several boosts. First, the channel length is shorter and increases the current drive of a device. Second, as the channel length gets shorter, the threshold voltage decreases due to the short channel effect and again increases the current drive capabilities. Third, as the channel gets shorter, the input capacitance is decreased due to less area across the gate oxide. In SOI, it has been mentioned that the there is a reduced short channel effect. Therefore, as the channel length is shortened in SOI, the improvement in current drive of a device is not increased as fast as a comparable bulk technology.

> **Rule of thumb:** Chips designed in PD-SOI are likely to exhibit 2/3 of bulk-CMOS' delay variability per standard deviation of channel length variation. For example, if a bulk CMOS design sees 6% delay change (off nominal delay) per sigma of channel length, its SOI counterpart will only see about 4% delay change per standard deviation of channel length.

8.7 SOI Timing Variability

Providing timing margin in high speed logic designs has typically required the designer to anticipate the impact on performance of (a) process variation, (b) activity-based device and interconnect wear-out, and (c) charging/discharging from preceding operations. Although temporal effects[3] (b and c above) contributed to delay variability in bulk CMOS, spatial effects[4] accounted for the vast majority. Delay variation is a concern in designs because the

Paths of different composition have different variability

Variability does not apply uniformly, even to paths of identical composition

Variability can not be anticipated in the timing

Providing margin to accommodate possible tolerance impacts performance

8.7.1 Total Delay Variation

In PD-SOI, temporal delay variation arising from changes in device body potential tends to assert timing tolerance roughly equivalent in magnitude to the spatial tolerance previously known to the industry [8.5]. TABLE 8-1 summaries all common sources of variation which might now be found in SOI chips.

Process/Technology	**Usage History**
Across-Chip Linewidth Variation	Self Heating (SOI)
Short Channel Effect Envelope	Floating Body (SOI)
Narrow Channel Effect Envelope	Residual Junction Charge
NFET/PFET Tracking: L_{EFF}, V_T	Hot Electron Effects
Across-Chip Interconnect Variation	Electromigration
Design	**Modeling**
Delay: History Effect (SOI)	Device Model Fit Error
Simultaneous Switching	Input Inversion Cap Onset
Noise: Charge Sharing	Timing Tool vs. Real Ckt. Response
Noise: Coupling, Supply, T-Line	
Across-Chip Temperature Variation	
Across-Chip Supply Variation	
Clock Skew	
Circuit Slew Rates	

TABLE 8-1 Sources of delay variation in PD-SOI CMOS

Rule of thumb: Delay variation caused by SOI's floating body effect is roughly equivalent to that change caused by variation in poly gate line width across a given chip.

---

3. *Temporal* describes mechanisms introducing variation which has a dependence on time.

4. *Spatial* describes mechanisms introducing variation which has a dependence on position/location

Delay variation, of course, is specific to the process implementation of the technology, and will vary from process technology to process technology. Nonetheless, Figure 8.16 illustrates how 100% of a path's total delay variation may be apportioned in a current PD-SOI technology microprocessor critical path.

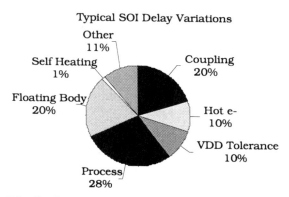

FIGURE 8.16 Distribution of the on chip delay variability for SOI.

8.7.2 History Variation

The variability associated with changes in body potential is in fact not constant. It follows that since history-dependent variability arises from changes its use history, any factors which globally influence body potential limits also affects history. Process, voltage, and temperature influence the range of voltages the body may assume.

Since the first switch is dependent upon the DC equilibrium of the body voltage and the body voltage is dependent upon the ratio of the drain to body diode's forward-bias current to the reverse saturation current of the source to body diode, any method to maintain that ratio will reduce the amount of variation in the delay during the first switch of the circuit. Such reproducibility requires that the implants of the source and drain junctions be consistent along with the thermal budget for the rest of the wafer processing.

The second switch is dependent upon the ratio of coupling capacitance to the body. One problem here is that the amount of charge injected into the body of a device is dependent upon the dV/dt of the gate and drain. Therefore, when the process does accurately reproduce the gate to body capacitor as well as the drain to body capacitance, the circuit design still influences the movement of the body.

Consider two identical inverters that have the same sizes, same process, and same input slew rate on the gate. If one inverter is more heavily loaded on its output, then the slew rate on its drain will be slower resulting in less charge being driven into the body of the NFET. This will raise the potential of the body less. Therefore, the more heavily loaded inverter will have a second switch that is slower than the second switch of a lightly loaded inverter.

One cannot ask that each circuit have the same output slew to have a constant second switch. First, no designer would agree to such a constraint. It would waste time and area. Second, every circuit has a load that varies with time. The load may change due to the switching of the down stream circuits or because a passgate was opened and now a larger capacitance is seen through the open passgate than before. This reduction in speed of the second switch can again cause the second switch to become slower than the first switch.

8.7.3 Body Initialization.

When establishing delay rules, the designer must establish the worst case and best case delays to be used in *Late Mode* and *Early Mode (or fast path)* timing, respectively. It is far too optimistic or pessimistic to assert the first switch or second switch delay for these rules, as a given circuit seldom is allowed to achieve these extremes. A common approach is to anticipate the range of body voltages by examining the activity factor for various circuits, assuming a given instruction and data stream. Given this range, a probability distribution may be derived, from which the chip designer may select a a body voltage to use in simulation which provides the desired comfortable level of design security. The distribution for a high speed processor is shown in Figure 8.17. This selection is not unlike the selection of product reliability, except that in this case, the probability of a whole path assuming that worst case history is quite remote. This works well for late mode timing. If the fastest case history was used on all devices, then the timing would be optimistic and unrealistic. It is not possible to set the body of all circuits in a given path to the fastest setting in hardware.

For early mode timing, it may be best to time with best case bodies. This eliminates the possibility of some very short path coming close to the best case history, since many fewer circuits are present in early mode paths.

Again, it is not possible to set up the body of every device to an accurate body potential for timing simulation. Therefore, one has to assume a range of possible frequencies for late mode and put in a bit of conservatism for early mode.

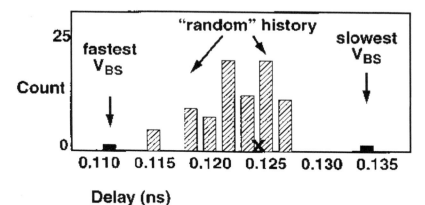

FIGURE 8.17 History Distribution for a high speed processor, plotting delay in ns (X axis) against number of paths possessing that delay. Final chip timing condition selection condition is marked by the X. [8.6]

8.8 Summary

Effective high speed logic design in SOI technologies requires vigilance not only in circuit design, but in the management of changes in chip global effects. These effects may arise from either the electrical properties of the SOI active elements, or from the physical properties associated with the SOI structure itself.

New temperature effects are present in normal use, in high temperature stress, as well as at lower temperature. In normal use, the heat associated with device activity can alter the transconductance of adjoining transistors. At stress conditions, elevated temperatures combined with the power consumption at elevated voltages can produce heat quicker than the package has the ability to dissipate. At low temperatures, the reduction in junction leakage results in history effects unlike those at higher use temperatures.

Changes in device capacitance induced by the presence of a buried oxide cause changes in noise immunity. Noise associated with power supply and lateral signal coupling may be substantially greater than that seen in bulk. Stronger relative SOI device strength exacerbates the problem device and must be accurately modeled. The device's higher transconductance also causes increased power consumption, which must be accommodated by the supply distribution and package cooling schemes.

Finally, because SOI provides MOSFET performance improvements, it is important for the global chip designer to remember that this technology does little to improve interconnect performance. Because of this, the actual performance improvement a given chip will enjoy is limited to the device delay associated with the worst cycle-limiting paths. In addition, the effects of both spatial and temporal delay variation must be budgeted in the selection of prudent product performance sort points.

REFERENCES

[8.1]   K. Bernstein, "How Low Can We Go?", *IEEE 1998 International Solid State Circuits Conference Evening Panel Discussion, "Will Power Limit High End Processor Performance?"* San Francisco, California, February 8, 1998.

[8.2]   *IEEE Transactions on Electron Devices*, August 1989.

[8.3]   I. Aller, et al., "SOI Steady State Temperature Dependence", *Proceedings of 1999 IEEE International SOI Conference,* October 1999, pp.100-101.

[8.4]   Kerry Bernstein, Keith Carrig, Christopher Durham, Patrick Hansen, David Hogenmiller, Edward Nowak, and Norman Rohrer, *High Speed CMOS Design Styles,* Boston, Kluwer Academic Publishers, 1998.

[8.5]   Sani R. Nassif, "Within-Chip Variability Analysis," *IEEE International Electron Devices Meeting, Technical Digest,* December 1998, pp. 283-286.

[8.6]   D. H. Allen, et al., "A 0.2μm 1.8V SOI 550MHz 64b PowerPC Microprocessor with Copper Interconnects," *Proceedings of 1999 IEEE International Solid State Circuits Conference,* February, 1999, pp. 438-439.

CHAPTER 9 *Future Opportunities in SOI*

9.1 Introduction

<u>If you can't beat'em, join 'em!</u>

The description of Partially Depleted SOI in the preceeding 8 chapters has hopefully equipped the reader with the concepts needed to anticipate and avoid the liabilities of floating body MOSFET devices. A number of alternatives, aimed at either effectively suppressing or exploiting the floating body for performance or reliability, are explored in this chapter. If SOI is a leading edge technology, then indeed, design methodologies which harness the floating body are farther out, and so are even less mature. Nonetheless, it is the strong belief of the authors that SOI's ultimate promise is realized in the achievement of these innovative topologies, and their application to high speed logic.

9.2 Floating Body Effect Suppression

Charge-balanced Device Design

Field Shield Isolation

BESS

Linked-Body Device

Patterned Insulators

As frequently noted, the SOI device's floating body creates design complexity. Although means have been described to accommodate the effect, it may be desirable to either globally or selectively suppress the floating body to assure robust design.

Globally, floating body suppression may be necessary to assure functionality of self-timed paths, or to guarantee satisfaction of a particular race condition. Frequently, margin must be added to cover delay variability in a half-cycle path[1], doubling the penalty on the chip's clock speed.

Locally, certain circuit structures require close tracking and tolerance of electrical properties in order to accurately time or evaluate activities. Timing exposures caused by floating bodies are found in additional skew caused by local clock regenerator variability, and FETs used as analog elements in phase-locked-loops based on charge-pumped voltage controlled oscillators [9.1]. Evaluation exposures include the classic cross-coupled differential sense amplifier design found with variation in most SRAM cache designs, and shown in Figure 6.4, "SRAM's Differential, cross-coupled sense amplifier, used in reading.," on page 124. Body contacts have already been discussed as a means of suppressing the floating body effect and will not be discussed further in this chapter.

9.2.1 Charge-Balanced Device Design

In Chapter 4, it was shown that the body voltage and threshold of a PDSOI MOSFET is influenced strongly by the device's body capacitance profile. Minimization of threshold voltage variation, and hence suppression of body history effects can be achieved to some measure by device design which balances charge distribution in different logic states and which contains non-ideal diode behavior [9.2].

---

1. "Half-cycle path" refers to a logic net with little margin, launched by a first clock and captured by a second clock in a chip L1/L2 timing scheme.

A goal of effective PDSOI device design is to minimize the deviation of initial body voltage from the device's stability condition to its steady state switching condition. Roughly speaking, the physical "knobs" defining the device's history characteristics are

1. the device extension's vertical depth,
2. the active silicon thickness, and
3. the thickness of gate's spacer

at a fixed channel length and operating voltage.

The AC contribution to reducing ΔV_T is reducing the gate-body capacitance in reference to body-source/drain capacitance. Shorter channel lengths, retrograde channel implants, and increased source/drain halo implant dose support this trend with scaling. Figure 9.1 below profiles the body charge distribution for an NMOS in its off-state and its on-state. Threshold variation can be minimized if

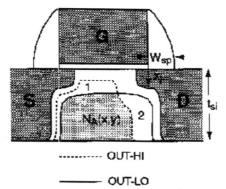

---------- OUT-HI

———— OUT-LO

FIGURE 9.1 Body charge balance in an NMOSFET

the charge contained in the depletion region to the gate when the gate is off is roughly equivalent to the charge contained in the depletion region facing the drain when the device is on.

The DC mechanism which has the ability to reduce history is diode non-ideality. Impact ionization in the off state creates hole charge accumulation in the body region. The more ideal the diode, the higher the body voltage will rise, and the higher the device's subthreshold current. Introducing additional diode leakage reduces the threshold voltage the device requires to achieve a fixed I_{off} spec in

---

the case of an inverter, where the source of the device is tied to GND.

In the upper device of a NAND stack, and in pass-gates when the source of the device may be at a higher potential, the body will charge, lowering the transistor's threshold voltage. In this scenario, junction leakage is not as important a parameter in determining performance. In a stack where the lower device is on causing the source of the upper device to be low however, junction leakage again can have a significant impact on performance.

9.2.2 Field Shield Isolation

Another novel means of mitigating floating body effect bounds the MOSFET's source, channel, and drain region with off-transistors rather than shallow trenches [9.4]. In Field Shield (FS) isolation technology, BOX formation is followed by conventional LOCOS or STI[2] device isolation, used to separate the body connections of adjacent devices. A field shield gate defining the actual device boundary is made, followed by more oxide and the formation of the desired transfer gate. A cross-section perspective of the resulting structure is shown in Figure 9.2 below. The field shield

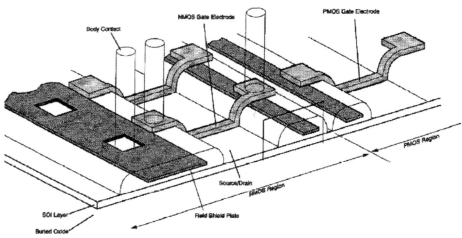

FIGURE 9.2 Field-Shield Device Structure [9.5]

2. Local Oxidation of Silicon, the defacto means of device electrical isolation preceeding Shallow Trench Isolation (STI) technology.

electrically bounds the diffusion regions with an "off" device, and in addition shields the device's channel sides from the influence of the active gate. Body voltage control is established from the sides, as holes accumulating in the body will flow to the contact connection made to the active silicon under the field shield device. Hence the device behaves essentially like a bulk CMOS device, and the noticeable "kink" in the SOI device's I-V curve caused by impact ionization is gone. The resulting MOSFET retains SOI's reduced substrate bias sensitivity. Because under reverse bias the body of the FS device is again electrically isolated, the body fully depletes, limiting body bias and fixing thresholds. Finally, FS device body resistance does not limit maximum switching speed, given the reduced body discharge scenarios associated with FS structures [9.5]. FS technology has been shown to be practical in gate array and high speed microprocessor applications.

Field shield devices require field shield gate bias connections and separate body region isolation, in addition to the body terminal needed in other body connection schemes. The parasitic coupling between the two gates must also be considered. The FS devices also require more area than traditional LOCOS or STI.

9.2.3 BESS

Bipolar-embedded source structures (BESS), introduced in 1996, fully suppresses partially-depleted SOI floating body effects by providing a recombination center for holes appearing in the body [9.6]. An example BESS structure incorporated in an NFET device is shown in Figure 9.3. Referring to the figure, a p-type recombination

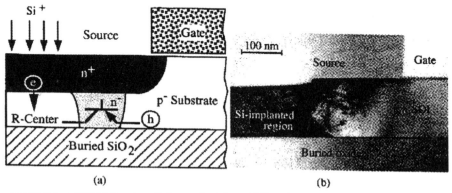

FIGURE 9.3 BESS Device: Schematic (a) and TEM image (b).[9.6]

center is established via implantation deep in the bottom of the source/drain diffusion at point R, away from the active device area. A passive bipolar transistor formed with an **n⁻** implant in the shunt region (base), provides a diffusion path for holes generated in the body (emitter) and this implant (collector).

Drain-Induced Barrier Lowering (DIBL), the drain voltage's inducement of short channel effect, is dramatically reduced in SOI MOSFETs when BESS structures are present, as the depletion region surrounding the source is reduced. The trade-off for this device enhancement is the added fabrication complexity and yield loss associated with the placement of the recombination center.

9.2.4 Body-Linked Device

Body-linked devices provide a means of eliminating the charge build up in the body of the FET while still providing reduced capacitance on the source and drain junctions [9.7]. Figure 9.4 provides a cross sectional view of the device along the width and

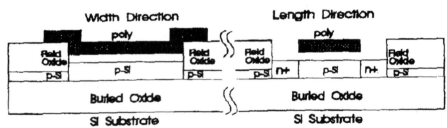

FIGURE 9.4 Cross section of linked body SOI FET structure [9.7]

length of the device. The key to this device is the incomplete field oxidation for the shallow trench isolation. This allows for the bodies of the devices to be resistively linked to each other. The on currents in this structure can be made similar, but the off current is substantially lower for the linked body device. The suppression of the floating body eliminates the kink effect and hence, V_T does not reduce as a function of drain voltage. Since a buried oxide is still present under the device, the area component of the junction capacitance is removed and the capacitance contribution to performance is preserved. Finally, with a linked body, the breakdown voltage of the FET is greatly reduced since the parasitic bipolar transistor is not able to turn on.

The intent of the approach above, as well as in Partial Trench Isolation (PTI) [9.8] is to provide regular bias contacts to external diffusions near the device. This diffusion then provides the desired body potential by contacting the body of a given device under its isolation trench, which does not reach all the way to the body. A cross-sectional diagram and SEM of the structure is shown in Figure 9.5. The spacing of these

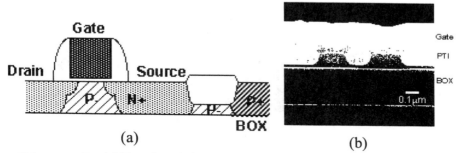

(a)

(b)

FIGURE 9.5 Partial Trench Isolation, cross-sectional diagram (a); and SEM (b) [9.8].

contacts must insure that the maximum resistance between the bias rail and the body does not exceed a given amount. This resistance in series with the cumulative body capacitance establishes the RC delay of the body tie, and hence the resulting history effect magnitude the device will still realize. In the referenced paper, 100 Kohms was used as the representative value, which roughly implies that a bias connection would be necessary approximately every 100μ. This bias may be fixed in potential, or could be varied as a function of desired standby current or temperature. Ostensibly, it could also be clocked to match specific machine states.

9.2.5 Patterned Insulators

The body of an SOI device may also be contacted from underneath, by electrically coupling the body to the substrate. Replacing the blanket SIMOX layer with a patterned oxide has been frequently proposed as a means of reducing junction capacitance without introducing floating body effects [9.9]. Figure 9.6 shows a cross-sectional diagram of such a structure, which provides a path from the body to the substrate, while still placing the area component of the source and drain regions upon a buried oxide.

From a circuit design point of view, patterned oxides are the perfect body ties. They take up no area, and further inhibit latch-up, while still asserting a known body potential. To the device designer, however, asserting a pattern on the gross SIMOX implant presents a multiplicity of challenges.

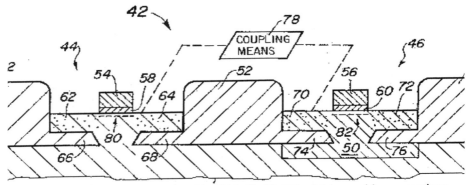

FIGURE 9.6 Patterned buried oxides (66, 68, 74, 76) which provide connections from the NFET bodies (80) and PFET bodies (82) to the underlying substrate or well (50) [9.9].

1. SiO2, as it is formed, grows in volume dramatically. This swelling places stress on the silicon lattice in the region of operation, changing device transconductance, and inducing fractures, pipes, and other defects in the device. In addition, the pattern's presence can induce deformities in the gate structure of the device.

2. The gross SIMOX implant must be perfectly aligned to the gate of the device, and its size must be controlled precisely, to achieve the desired electrical characteristics. This precision is not easily obtained in an implant as massive as SIMOX.

3. The source and drain diffusion implant dose and energy needs to track the final position of the insulator, so that given the devices thermal budget and implant outdiffusion, the resulting junctions remain on top of BOX.

While patterned oxides in theory would be perfect body connections, in practice their construction presents substantial challenges to high quality volume production processing.

9.3 DTCMOS

DTCMOS Low Voltage Static Logic
0.5V Pass Gate Logic
DTCMOS Elevated Voltage Schemes

9.3.1 DTCMOS Low Voltage Static Logic

Dynamic Threshold CMOS (DTCMOS) is the technique of contacting the polysilicon gate of an SOI device to its electrically isolated body region, outside of the active channel region [9.10]. By driving the body potential to the same voltage as the gate, the device is no longer tied to the attributes of a single threshold. In Figure 9.7, the

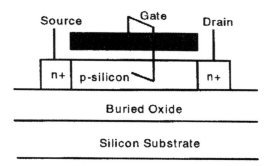

FIGURE 9.7 Dynamic Threshold CMOS (DTCMOS), tying body to gate [9.10]

gate of the device is tied through a buried contact to its body. This configuration has a number of advantages.

- **"ON" PERFORMANCE**
  When the gate voltage of the DTCMOS device is driven high, the device's body potential rises as well, reducing the change in potential required to create an inversion region. In effect, the body of the device is employed as a "back-gate," assisting in the inversion of the channel region. This presents as a lower effective device threshold voltage (in the on-state) than which would have been realized had the device been built in bulk CMOS or SOI. If the body had been left floating as in conventional SOI, the low "on" threshold voltage which DTCMOS achieves

would be not be realized. In the floating body case, the body begins high, but discharges through the source junction as the source falls, immediately elevating the threshold voltage.

- **"OFF" LEAKAGE**
  When the gate voltage of the DTCMOS device is driven to GND potential, the body potential falls as well. This increases the potential required to create the inversion region, causing the higher effective device threshold voltage (in the off-state) associated with a device built in bulk CMOS. The advantage here is that in the off state, the device enjoys lower leakage currents. If the body had been left floating as in conventional SOI, the probability of non-zero body bias is higher, and with it higher leakage and noise immunity degradation.

At first glance, then, it would appear that DTCMOS device incorporates the best attributes of a low threshold voltage device when on, and of a high threshold voltage device when off. This benefit, however, must be weighed against the following liabilities, unique to the specific process definition:

- **VOLTAGE**
  DTCMOS devices must be operated at voltages of 0.5 V or less, so that the gate connection to the body, when high, can not appreciably forward bias the body-source or body-drain diode junctions. This operating point severely degrades the transconductance of the device and increases the current consumption.

- **LOAD**
  The additional gate load presented by body's capacitance to both source and drain, gate, gate overlap, and inversion channel increases the overall switching delay of the device. Depending on device composition, the added delay may exceed the improvement in I_{DSAT}.

- **MILLER EFFECT**
  On the front device of the conventional MOSFET, the overlap capacitance comprising the device's Miller response is a substantially lower proportion of the total gate capacitance, and so feed-forward effects are manageable. In addition, that capacitance is through an insulator (SiO_2) with substantially lower dielectric constant than silicon. On the back device, however, the source and drain junction capacitance to the body is quite large with respect to the body-inversion layer capacitance when the device is on, and through a capacitor with the same dielectric constant [9.11]. The Miller response is very expensive for DTCMOS.

- **RC DELAY**
  Since the back gate (body) of the device is not silicided like the polysilicon front gate, the lateral resistance through which the gating potential must propagate is substantially higher. Attempts to reduce its resistance by thickening the active sili-

con layer only increase the body-to-source and body-to-drain capacitance, exaggerating the Miller response described above. High gate RC reduces the maximum F_T the device can provide.

The power consumption of DTCMOS may be larger due to several components. First, at higher voltage, the DC power becomes exponentially larger as the voltage is raised, due to the forward biased diode from the body to source. Secondly, the capacitance is larger which increases the amount of charge that has to be shuttled during transitions. Finally, DTCMOS offers lower threshold voltages, so the resulting crossover current can be larger. However, the delay of a DTCMOS inverter has been shown to be reduced, creating a smaller power-delay product at voltages less than 0.7V [9.16] . The speed improvement is dependent upon the output load of the inverter. If the inverter is unloaded, then DTCMOS is slower than normal SOI, but if the load is substantial, the delay improvement can be larger than that offered by PD SOI [9.17] .

Clearly, the advantages of DTCMOS depends upon the specific implementation. The above issues should be considered when evaluating a particular device.

9.3.2 0.5V Pass-Gate Logic

The device structure highlighted in Section 9.3.1 has been used to implement a number of logic functions described in the literature. A noteworthy example is the implementation of "gate-body connections" (GBC) in 0.5V SOI CMOS [9.12]. The referenced paper reported a buffer chain using the structure shown in Figure 9.8 reduced its delay to 33% of that of the conventional SOI CPL running at 0.5V. In this

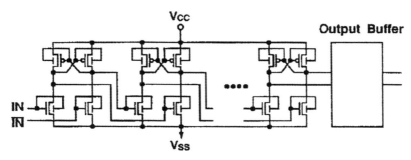

FIGURE 9.8 GBC Scheme buffer chain in 0.5V SOI CMOS. [9.12]

topology, the gate input of each transistor is also coupled to the body of its device.

The pull-up PFET device, when turned on retains a high threshold voltage until the output node is pulled sufficiently low. This follows as a result of the body being tied directly to the output node, and causes added short circuit current.

A more effective implementation is the "input body connection" (IBC) configuration shown in Figure 9.9.. With this arrangement, the gates of the PFETs are cross-cou-

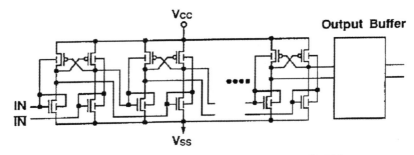

FIGURE 9.9 IBC Scheme buffer chain in 0.5V SOI CMOS. [9.12]

pled, but the bodies are not. In fact, both NFET and PFET bodies on each side are tied directly to the respective inputs. The threshold voltage of the turned-on PFET falls immediately, as its body is tied to the input rather than output. In this case, the literature reports an added improvement of 36% over the GBC topology at 0.5V

9.3.3 DTCMOS Elevated Voltage Schemes

Among the discouraging features of DTCMOS which inhibit industry acceptance (See page 204), the worst is the requirement that V_{DD} be at most equal to V_{DIODE}, approximately 0.5V. Novel approaches have been proposed for enjoying DTCMOS action while retaining higher voltage operation, compatible with other system components [9.13]. A particularly innovative approach is shown in Figure 9.10. This topology comprises a conventional DTCMOS schematic altered by the addition of a back-gate biasing circuits. The back-gate circuits consist of capacitors and resistive loads which forward-bias the bodies with respect to the sources. By providing bias to the bodies of the inverter devices, higher performance and lower standby currents are realized. When input Vin transitions high, the voltages of the bodies of devices M_1 and M_2 are also elevated in voltage, elevating the threshold voltage of M_1 and depressing the threshold voltage of devices M_2. Since capacitor C1 and C2 are in

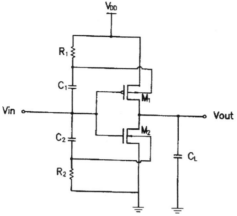

FIGURE 9.10 Dynamic Threshold Voltage Scheme for $V_{DD} > V_{DIODE}$ [9.13]

series between, Vin and the body of the devices, the voltage swing on Vin is not limited to the diode drop. The coupling across the capacitor conditions the devices in the desired direction, as the rising input seeks to pull down the output. After the transition has been secured, the next design goal is to stem the leakages associated with low threshold voltages, and recondition the bodies for the next transition. This is achieved via resistors R_1 and R_2, which return the body and capacitor voltages to V_{DD} and GND potentials, respectively. SOI enables this approach by providing an electrically isolated body which can sustain independent biasing.

A second method for making gate connections to the body, which allows for an increase in the supply voltage, is called Transistor Coupled Body [9.18] and is shown in Figure 9.11. Here, a second small transistor is built near the primary transistor and shares a low resistance connection to the body and a highly resistive connection to the source. The body of the primary transistor is controlled by the second transistor, which switches in parallel with the primary transistor. When the input, V_{in}, is high, the body of the primary transistor is pulled high (when V_{out} starts high) as the second transistor turns on. When the input is low, the body of the primary transistor is pulled to GND through a large resistance and the leakage current is very small. The features of DTCMOS are therefore realized, establishing a large Ion/Ioff ratio over a wide voltage range.

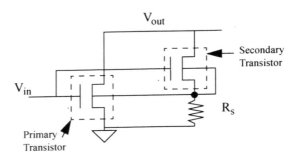

FIGURE 9.11 Transistor Coupled Body for DTCMOS at elevated V_{DD}.

9.4 DGCMOS

Dual Gate CMOS (DGCMOS) is an example of a larger family of device structures which may be loosely considered an SOI enhancement. The concept is discussed here if only because of the buried oxide insulator it has in common with the PD-SOI CMOS device described throughout this text.

Figure 9.12 provides a simplified side view of the DGCMOS device as proposed in the referenced literature. DGCMOS provides a means of actively driving the MOS-

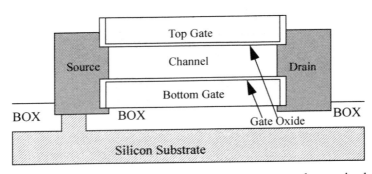

FIGURE 9.12 Side View, Double-gate MOSFET as recently practiced.

FET mode of operation via contacted gates placed both above and below the conducting channel [9.14]. In partially-depleted SOI, contact to the body provided a means of controlling the "back-gate" of the transistor. That approach had disappointing characteristics:

1. The body control potential needed to remain within one V_{DIODE} of the source in order to not sink large amounts of forward bias diode current. For that reason, the control could not conventionally come from the gate. DGCMOS is free to drive the bottom-gate to the same top-gate potential.

2. A back-gate formed by the body region has exceedingly high resistance, limiting the RC response of such a connection. In DGCMOS, the bottom gate ostensibly could be engineered for lower resistivity.

3. The parasitic capacitance of the back gate to the source and drain junctions in the body-contacted case can be quite high, embodied in effectively Miller capacitance formed between the drain and body (a "back gate" in this case). In the DGCMOS proposed structure, the back capacitance is self-aligned to the channel in much the same way as the front gate, producing similar, manageable parasitics.

4. The junction interface of the PD-SOI back gate to the source and drain regions permits the presence of the parasitic bipolar device. In DGCMOS, the gate oxide prevents this device from forming.

5. The most important advantage is the effective doubling of the channel width. Although the device is now essentially a fully-depleted structure, inversion channels form most strongly against the front and back gates, increasing the I_{DSAT} of the device.

New process technology provides a means of opening a window in the BOX to establish a seed from which selective epitaxial silicon may be grown. The silicon is grown through a tunnel in the gate silicon established with amorphous silicon. The grown silicon extends up from the seed layer forming the source, through the tunnel and contacting the drain silicon.

DGCMOS may provide larger ratio of Ion/Ioff than other SOI structures [9.15] .

9.5 3-Dimensional SOI

The observation that floating body SOI MOSFETs no longer need to be electrically associated with a substrate has not been lost on a number of researchers. It has been argued that, if the means exist to selectively grow an epitaxial silicon layer on top on an insulating layer, (say the inter-layer-dielectric conventionally used to isolate multiple wiring layers from one another!), then the designer is no longer constrained to one active element monolayer [9.21]. Recent published work suggests that performance improvement is achieved predominantly by shorter interconnects in such a structure, and is optimized when 2 device layer structures are formed and used. Interconnect performance improvements of up to 18% might be expected with the resulting shorter wire. A proposed structure is shown in Figure 9.13.

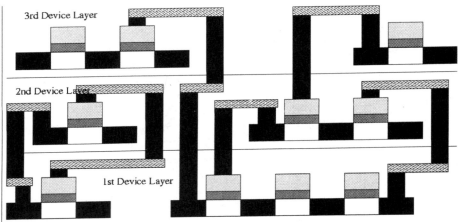

FIGURE 9.13 Cross-section of 3D integrated circuit [9.21].

An intriguing idea, to be sure! But one which will take our industry a while to embrace. The normal boundary conditions for power/heat dissipation, process complexity and cost, and realized performance benefits are formidable in such a structure, and will require more development. It will be fun watching its development, however.

9.6 Future Scaling Opportunities

The justification for investing in a new CMOS device technology such as SOI must be founded not only on it's present attributes, but on its ability to favorably scale those features through future generations of lithography and voltage. Bulk technologies are realizing diminishing returns from scaling, due in part to the inability to keep the ratio of threshold voltage to supply voltage constant.

In SOI, floating body effects at higher drain voltages induce increased leakage. To counter this leakage, default thresholds are increased, leading some to the conclusion that SOI suffers the same ultimate fate with scaling as its bulk predecessor [9.19]. It has been shown more recently, however, that as the supply voltage drops, threshold voltage may again be scaled more aggressively, as impact ionization is substantially reduced at lower voltages [9.20]. The difference between a MOSFET's V_{TLIN} and V_{Tsat} is an effective means of assessing the impact of supply voltage on barrier height. Figure 9.14 demonstrates a noticeable reduction in drain-induced barrier lowering (DIBL) with voltage for the scaled SOI device, suggesting that in fact lower thresholds will be acceptable in future SOI generations. The comparable bulk device shows DIBL increasing to unacceptable levels below channel lengths of about 0.08μ.

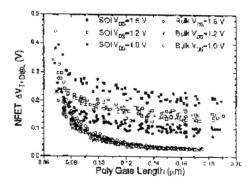

FIGURE 9.14 Difference between linear and saturated threshold voltage, as a function of gate length and supply voltage for bulk and SOI NFETs.

The more important result may be the subsequent effect on subthreshold currents. With the strong impact ionization dependence on V_{DS}, it follows then that standby currents should fall much more precipitously in SOI with scaling. As shown in Figure 9.15, standby currents are higher than that found in a comparable bulk MOSFET at elevated V_{DS}. But for lower supply voltages at shorter channels, however, SOI leakage currents are consistently less than that observed in bulk devices. The data suggests this crossover occurs at V_{DS} of 0.7 V or less.

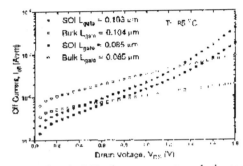

FIGURE 9.15 NFET subthreshold leakage current vs drain-source voltage.

9.7 Summary

As we began in this text, we note the pressures of scaling have made it increasingly difficult to live with the electrical characteristics of a single, static device. With a common substrate, one is asked to accept standby currents that are too high, or performance which is disappointing. Isolating the body of the transistor allows the designer to finally realize the fourth terminal of the MOSFET. The state of the art in PD-SOI, unfortunately, has only begun to evolve more elegant solutions for contacting and controlling the bias on this fourth terminal. Because of this deficiency, the industry has been motivated to understand the behavior of this node if left floating, so that the penalties of body contacts may be avoided. Solutions such as Field Shield, BESS, linked body and PTI have been proposed as longer term solutions and were described. Other future approaches which integrate body bias into the logic, such as DTCMOS promise more dynamic control, but have shortcomings which need to be addressed.

So, will SOI become predominant in our industry? Perhaps not exactly in its present incarnation. But as in "The Firebird," in which one feather ultimately helps the prince free himself of the evil sorcerer, a simple buried oxide layer will likely remain a powerful means of extending CMOS technology past the immediate scaling barriers at hand, and into yet more generations of lithography.

REFERENCES

[9.1]   F. M. Gardner, *"Phaselock Techniques,"* Wiley Publishing, ISBN 0471042943, 1979

[9.2]   A. Wei, et al., "Design Methodology for Minimizing Hysteretic Vt-Variation in Partially-Depleted SOI CMOS," *Proceedings of IEDM 1997*, pp. 411-414.

[9.3]   J. Eckhardt, et al., "A SOI-Specific PLL for 1 GHz Microprocessors in 0.25 μm 1.8V CMOS," *Proceedings of 1999 IEEE International Solid State Circuits Conference*, February 1999, pp. 436-437.

[9.4]   T. Iwamatsu, et al., "CAD-Compatible High-Speed CMOS/SIMOX Gate Array Using Field-Shield Isolation," *IEEE Transactions on Electron Devices*, Vol. 42, No. 11, November 1995, pp. 1934-1939.

[9.5]   S. Maeda, et al., "Suppression of Delay Time Instability on Frequency using Field Shield Isolation Technology for Deep Sub-Micron SOI Circuits," *Proceedings of IEDM 1996,* December 1996, pp. 129-132.

[9.6]   M. Horiuchi, et al., "BESS: A Source Structure that Fully Supresses the Floating Body Effects in SOI CMOSFETs," *Proceedings of IEDM 1996*, pp. 121-124.

[9.7]   W. Chen, Y. Taur, D. Sadana, K.A. Jenkins, J. Sun, and S. Cohen, "Suppression of the SOI Floating-body Effect by Linked-body Device Structure", *1996 Symposium on VLSI Technology Digest of Technical Papers,* June 1996, pp. 92-93.

[9.8]   Y. Hirano, et al., "Bulk-Layout-Compatible 0.18μ SOI-CMOS Technology Using Body-Fixed Partial Trench Isolation (PTI)", *1999 IEEE International SOI Conference*, October 1999, pp. 131-132

[9.9]   T. Kamins, et al., "Method For Making Patterned Implanted Buried Oxide Transistors and Structures", *U.S. Patent Number 4,810,664*, March 7, 1989

[9.10] F. Assaderaghi, et al., "A Dynamic Threshold Voltage MOSFET (DTMOS) for Ultra-low Voltage Operation," *IEDM Technical Digest*, pp. 809-812, December 1994.

[9.11]   C. Wann, et al., "Channel Profile Optimization and Device Design for Low-Power High-Performance Dynamic-Threshold MOSFET," *Proceedings of IEDM '96*, Paper 5.3, pp. 113-116

[9.12]   T. Fuse, et al., "0.5V SOI CMOS Pass-Gate Logic," *ISSCC Digest of Technical Papers*, February 1996, pp. 88-89.

[9.13]   Chen; Ming-Jer, et al., "Dynamic threshold voltage scheme for low voltage CMOS inverter," *U.S. Patent 5,644,266*. July 1, 1997

[9.14]   H. S. P. Wong, et al., "Self-Aligned (Top and Bottom) Double-Gate MOSFET with a 25 nm Thick Silicon Channel," *Proceedings of 1997 International Electron Devices Meeting (IEDM)*, pp. 427-430.

[9.15]   Liqiong Wei, Zhanping Chen, and Kaushik Roy, "Double Gate Threshold Voltage (DGDT) SOI MOSFETs for Low Power High Performance Designs", *Proceedings of 1997 IEEE International SOI Conference*, October 1997, pp. 82-83.

[9.16]   Wei Jin, Philip C. H. Chan, and Mansun Chan, "On the power Dissipation in Dynamic Threshold Silicon-on-Insulator CMOS Inverter", *IEEE Transactions on Electronic Devices*, Vol. 45, No. 8, August 1998, pp.1717-1724.

[9.17]   Glenn Workman and Jerry Fossum, "A Comparative Analysis of the Dynamic Behavior of BTG/SOI MOSFET's and Circuits with Distributed Body Resistance", *IEEE Transactions on Electronic Devices,* Vol. 45, No. 10, October 1998, pp. 2138-2145.

[9.18]   Masatada Horiuchi, "A Dynamic-Threshold SOI Device having an Embedded Resistor and a Merged Body-Bias-Control Transistor", *Technical Digest of the International Electron Devices Meeting 1998"*, December 1998, pp. 419-422.

[9.19]   R. Chou, et al. "Scalability of Partially Depleted SOI Technology for sub 0.25 μm Logic Applications," *Technical Digest of the International Electron Devices Meeting 1998"*, December 1998, pp. 591-594.

[9.20]   E. Leobandung, et al., "Scalability of SOI Technology into 0.13μm 1.2V CMOS Generation", *Technical Digest of the International Electron Devices Meeting 1998"*, December 1998, pp. 403-406.

[9.21]   R. Zhang, et al., "Architecture and Performance of 3-Dimensional Circuits," *Proceedings of 1999 IEEE International SOI Conference*, October 1999, pp. 44-45.

INDEX

217

R
read disturb 131
Read Leakage 124

S
Scaling 1
self heating 55, 181
Sense Amplifier Bias 124
SER 59
shallow trench isolation 14
Short Channel Effect 46, 68
silicon on sapphire 13
SIMOX 14, 18
Simultaneous Switching 178
Sleep current 180
Smart Cut 14, 17
Soft Errors in Dynamic Logic 113
space-charge region 30
SRAM 119
  bitline 119
  cell stability 131
  read disturb 131
  reading a cell 123
  sense amp mismatch 124
  write 120
steady state 40, 75
Substrate Hot Electron 34
Supply Rail Droop 135

T
Technology Scaling 1
thermal runaway 168, 182, 183
Timing Variability 187
transfer curve 79

V
Voltage Reference 156
Voltage Stress 108

About the Authors

Kerry Bernstein is a senior technical staff member and lead technologist in the Microprocessor Development Group at IBM Microelectronic Division's Essex Junction, Vermont facility. He is currently responsible for future product technology definition, performance and application. Mr. Bernstein received his engineering degree from Washington University in St. Louis, and joined IBM in 1978. He holds 20 US Patents, and is co-author of a popular textbook on high speed circuit design. His interests are in the areas of low power and high speed electronics. Mr. Bernstein is a senior member of the Institute of Electrical and Electronic Engineers.

Norman Rohrer is first and foremost a husband and a father to two boys. Norman's other credentials stem from receiving his Bachelor of Science in physics and mathematics from Manchester College, North Manchester, IN in 1987. He received his Master's Degree and Doctor of Philosophy degree in Electrical Engineering from The Ohio State University, Columbus, OH, in 1990 and 1992, respectively. Norman is a senior engineer in the PowerPC Microprocessor Development group at IBM Microelectronics in Essex Junction, VT. Norman holds 3 patents and is the co-author, along with Kerry, of the book *High Speed CMOS Circuit Design Styles*.